T0291523

Suzhou Industrial Park

Achievements, Challenges and Prospects

EAI Series on East Asia

ISSN: 2529-718X

About the Series

EAI Series on East Asia was initiated by the East Asian Institute (EAI) (http://www.eai.nus.edu.sg). EAI was set up in April 1997 as an autonomous research organisation under a statute of the National University of Singapore. The analyses in this series are by scholars who have spent years researching on their areas of interest in East Asia, primarily, China, Japan and South Korea, and in the realms of politics, economy, society and international relations.

Published:

Suzhou Industrial Park: Achievements, Challenges and Prospects
by John WONG and LYE Liang Fook

Chinese Society in the Xi Jinping Era
edited by ZHAO Litao and QI Dongtao

China's Economic Modernisation and Structural Changes: Essays in Honour of John Wong
edited by ZHENG Yongnian and Sarah Y TONG

Politics, Culture and Identities in East Asia: Integration and Division
edited by LAM Peng Er and LIM Tai Wei

China's Development: Social Investment and Challenges
by ZHAO Litao

China's Economy in Transformation under the New Normal
edited by Sarah Y TONG and WAN Jing

Contemporary South Korean Economy: Challenges and Prospects
by CHIANG Min-Hua

*The complete list of the published volumes in the series can also be found at https://www.worldscientific.com/series/eaisea

EAI Series on East Asia

Suzhou Industrial Park

Achievements, Challenges and Prospects

John WONG
East Asian Institute
National University of Singapore
Singapore

LYE Liang Fook
ISEAS-Yusof Ishak Institute, Singapore

NEW JERSEY • LONDON • SINGAPORE • BEIJING • SHANGHAI • HONG KONG • TAIPEI • CHENNAI • TOKYO

Published by

World Scientific Publishing Co. Pte. Ltd.
5 Toh Tuck Link, Singapore 596224
USA office: 27 Warren Street, Suite 401-402, Hackensack, NJ 07601
UK office: 57 Shelton Street, Covent Garden, London WC2H 9HE

Library of Congress Cataloging-in-Publication Data
Names: Wong, John, 1939– editor. | Lye, Liang Fook, editor.
Title: Suzhou Industrial Park : achievements, challenges and prospects /
edited by John Wong (East Asian Institute, National University of Singapore, Singapore),
Liang Fook Lye (ISEAS-Yusof Ishak Institute, Singapore).
Description: New Jersey : World Scientific, 2019. | Series: EAI series on East Asia
Identifiers: LCCN 2018055908 | ISBN 9789811200038
Subjects: LCSH: Suzhou gong ye yuan qu (Suzhou Shi, Jiangsu Sheng, China) |
Industrial districts--China--Suzhou (Jiangsu Sheng) | Suzhou (Jiangsu Sheng, China)--
Economic conditions. | Singapore--Foreign economic relations--China--Suzhou (Jiangsu Sheng) |
Suzhou (Jiangsu Sheng, China)--Foreign economic relations--Singapore.
Classification: LCC HC428.S8 S779 2019 | DDC 338.900951/136--dc23
LC record available at https://lccn.loc.gov/2018055908

British Library Cataloguing-in-Publication Data
A catalogue record for this book is available from the British Library.

For any available supplementary material, please visit
https://www.worldscientific.com/worldscibooks/10.1142/11269#t=suppl

Typeset by Stallion Press
Email: enquiries@stallionpress.com

Printed in Singapore

Contents

About the Authors

Professor John WONG was professorial fellow and academic adviser to the East Asian Institute (EAI) of the National University of Singapore and formerly research director of EAI, and director of the Institute of East Asian Political Economy (IEAPE). He taught Economics at the University of Hong Kong in 1966–70 and at the National University of Singapore in 1971–1990.

He had been short-term academic visitor at the Fairbank Centre of Harvard University, Economic Growth Centre of Yale University, St Antony College of Oxford University and Economics Department of Stanford University. He had held the ASEAN Chair at the University of Toronto.

He had written/edited 40 books, and published over 400 articles and papers on China, and development in East Asia and ASEAN. His first book was *Land Reform in the People's Republic of China* (New York, Praeger, 1973) and his most recent book was *Zhu Rongji and China's Economic Take-off* (London, Imperial College Press, 2016). In addition, he had written numerous policy-related reports on development in China for the Singapore government. He obtained his PhD from the University of London.

LYE Liang Fook is senior fellow of the Regional Security and Political Studies Programme at ISEAS — Yusof Ishak Institute whose research interests cover China-ASEAN relations as well as China-Singapore relations. As co-coordinator of the Vietnam Studies Programme, Liang

Fook is responsible for facilitating deeper research and empirical understanding on Vietnam's domestic situation as well as Vietnam's foreign policy.

He was a key member of a study team commissioned by the Singapore government to conduct a study of the Suzhou Industrial Park, the first government-to-government project between Singapore and China. He has also followed closely and written about developments on the Tianjin Eco-city and Chongqing Connectivity Initiative, the second and third flagship projects between the two countries.

Liang Fook previously managed the Singapore secretariat of the Network of East Asian Think Tanks (or NEAT for short) and the Network of ASEAN-China Think Tanks (or NACT for short), two Track II initiatives that have been approved at the ASEAN plus Three level and ASEAN plus One level respectively to promote regional cooperation. He was previously research fellow and assistant director at the East Asian Institute of the National University of Singapore where he completed the drafting of this publication.

Preface

By LYE Liang Fook

This publication on Suzhou Industrial Park is dedicated to the memory of Professor John Wong who unexpectedly passed away on 11 June 2018. The Suzhou Industrial Park had been an enduring area of Professor Wong's research focus since the early 2000s when the East Asian Institute (EAI) conducted an in-depth study to draw lessons of experience arising from the Singapore government's involvement in this project. Since then, Professor Wong had closely followed and written on developments in the Suzhou Industrial Park. Just before his demise, he was still mulling over the most suitable title for this publication as well as the chapter titles. He even instructed me to follow-through with this publication and my initial thought was that he was in his usual self of assigning work.

To me, Professor Wong has been a seminal figure at the East Asian Institute (EAI) of the National University of Singapore as well as in my life as a younger scholar doing research on China. At EAI, Professor Wong had left an indelible imprint and been instrumentally involved in various stages of EAI's development, first as director of the Institute of East Asian Philosophies (from late 1990 till 1992) and Institute of East Asian Political Economy (from 1992 to 1997), followed by research director of EAI (for a long stretch of 12 years from 1997 to 2009) and finally professorial fellow of EAI (from 2010 till 2018).

Professor Wong would often recount to me that he had not expected to land a job at the Institute of East Asian Philosophies, the predecessor of EAI. He told me that sometime in September 1990, he received a totally unanticipated phone call from Dr Goh Keng Swee's personal assistant for him to meet Dr Goh at his Monetary Authority of Singapore office. Dr Goh was chairman of the institute, vice chairman of the Monetary Authority of Singapore and economic adviser to China's State Council "Office of Special Economic Zones" for some years. After sizing him up, Dr Goh was of the view that Professor Wong was suitable as director of the institute as he was a development economist with a special interest in China and East Asia. Before taking up the offer, Professor Wong had consulted several of his colleagues who mostly cautioned him against making such a drastic move because they felt that it would be a mammoth challenge for an academic like him to work with Dr Goh. However, Professor Wong eventually took the plunge and would occasionally reminisce about his unexpected transition from academic to think tank director.

One of the key things that Professor Wong told me that he picked up under Dr Goh was writing policy briefs of interest for the Singapore government. Professor Wong said that Dr Goh instilled in him the importance of keeping the policy briefs concise, readable and informative as opposed to the usual approach of academics for long-winded narratives. Professor Wong would pay particular attention to how the policy briefs are written and presented. He even meticulously deliberated on what he thought would be the most appropriate main title and sub-titles for the policy briefs as well as the specific wording used. He would often remind me that even after one had completed the writing of a product and was satisfied with it, one should not rush to submit it. It would be better to set it aside, mull over it and then revisit it again. Following his sound advice, I would invariably find areas where I could improve on such as rewriting a sentence or replacing a word here and there. To him, a product could always be improved further, akin to how a rough jade could eventually be transformed into a precious gem after several rounds of rigorous polishing. Today, these policy briefs or what is commonly known as *Background Briefs* continue to be produced and circulated to the government. EAI would even hold meetings with relevant government agencies to better understand their focus and produce *Background Briefs* relevant to this focus.

At the personal level, I have benefitted greatly from the wise counsel of Professor Wong. Apart from instilling in me the skills and importance of writing well and in a concise manner as recounted earlier, Professor Wong stressed the importance of reading widely so that one is not merely confined to a particular area of specialization. To him, a mature scholar was one that was not short of ideas and not short of topics to write on. To him, reading widely and more importantly, reading beyond one's specific area of specialization, would enable one to see the bigger picture and how one piece of the puzzle could be related to another piece of the puzzle. By developing an interest in other topics or related topics, one would then be able to move beyond one's usual comfort zone and venture into new areas. In this way, a scholar would be able to develop new interests and new areas to write on, and over time expand his or her scope and depth of knowledge. Professor Wong would further urge his EAI colleagues to read beyond the normal office hours. To set an example, he would bring home reading materials from office and read it from the comfort of his home. To him, reading was not a chore or work but a passion. He often would exhort scholars to carry a bag so that he or she could bring home some reading materials.

Professor Wong was full of energy and extremely inquisitive. In my travels with him abroad, Professor Wong would enjoy exploring the sights and sounds of the cities. He loved to "walk the ground" and would come prepared with a pair of walking shoes, his sunshades and hat, and slinging along a canvas bag that contained the essentials of a traveller such as mineral water, pills for all sorts of ailments that one could possibly think of and even some snacks in case one would fall hungry while walking. Often, he would be the one blazing a trail while walking and I would be the one trailing him.

In his overseas travels, Professor Wong also liked to visit markets or supermarkets to understand the variety of products available and, more importantly, the pricing of each product so as to get a sense of the cost of living for the local residents. To him, this was an indicator of the quality of life and whether the local government was able to fulfil its basic function of ensuring that there was ready supply of daily goods and services for its people. He was truly an economist at heart and in practice! And while having meals at local restaurants, Professor Wong would make sure that no food would be wasted at the table. To him, having leftover food was an "unforgiveable sin". If there were any leftovers, he would take it upon himself to divide the leftovers among those travelling with him.

No amount of words can describe the loss that I experienced when I learnt of Professor Wong's sudden demise in June 2018. I had in fact accompanied him on an overseas working trip in early 2018 and he was his usual energetic self. Looking back, Professor Wong had enriched my life in the years I spent at EAI. I vividly remember that before I left EAI in early 2018, I had told him that I regarded him as a "fatherly figure" who took care of those under his wing. Although this wing is no longer around, I am confident that those who have "grown up" and benefitted from his wise counsel would have developed strong wings to carry on the good work that he had done when he was with EAI.

Introduction
Singapore-Suzhou Industrial Park 20 Years On: Development and Changes

John WONG and LYE Liang Fook*

The Suzhou Industrial Park (SIP), inaugurated in 1994 with the strong support from the governments of Singapore and China, will be celebrating its 25th anniversary on 12 June 2019. This quarter of a century anniversary will mark a major milestone in the ever-evolving Singapore-China relationship. From a project that has survived the many operational hiccups at the start-up phase to its present standing as an unrivalled industrial park in China, the SIP has indeed come a long way. Through collaborating on the SIP, not only have the businesses from both sides benefitted from the economic opportunities but also past, present and succeeding generations of Singapore and Chinese leaders have gotten to know each other better. The political and economic networks built up over the years have proved most invaluable. The SIP stands as an example of what can be jointly achieved once both sides set their hearts and minds to it. It is a concrete symbol of the substantive relationship between China and Singapore.

* John WONG was professorial fellow at the East Asian Institute (EAI) of the National University of Singapore and former research director of EAI and director of its predecessor the Institute of East Asian Political Economy. LYE Liang Fook is senior fellow of the Regional Security and Political Studies Programme at ISEAS — Yusof Ishak Institute.

No Easy Start

Back in 1994, as bulldozers were busy working on the two sq km site (the initial phase for a total of 70 sq. km)[1] of the flagship SIP project and filling the low-lying rice fields up to five metres high, Chinese engineers were hotly questioning their Singaporean counterparts if such a stringent and costly landfill would really be necessary. Every additional centimetre of landfill on such a scale would, of course, add dearly to the cost of the land, apart from the longer time it would have to take to lay the foundation. China's alternative strategies of relying on its traditional methods of draining the site by digging canals and building dams would be far cheaper and, in the view of Chinese engineers, equally effective.

The Singapore side firmly stood its ground. Having carefully studied the last severe flood that struck Suzhou in 1991, Singapore's experts came to the conclusion that the landfill of five metres, though expensive, would be needed to withstand serious potential floods in future. Slated to be a capital-intensive high-tech industrial park, SIP would have to be completely safe from flooding in order to protect expensive equipment and gain the confidence of foreign investors. Singapore planners therefore took no chances. Five years thereafter, in 1999, as the Yangtze River valley was hit by the "worse floods in 100 years", the factories in SIP were spared while its adjacent areas were ravaged by flooding. Singapore's points of view were thus quietly vindicated.

However, Singaporeans were not in Suzhou to win arguments with Chinese officials. They were there to do a good job, to build a solid infrastructure for a world-class industrial park based on Singapore's past success with its Jurong Industrial Estate.[2] It is not a matter of

[1] The size of the Singapore-China cooperation zone was initially 70 sq km. It was announced in August 2006 that this cooperation zone would be expanded by an additional 10 sq km, making it 80 sq km.

[2] Efforts to clear and prepare the land for the Jurong Industrial Estate started in 1961 under the aegis of the Economic Development Board. See Jurong Heritage Trail by National Heritage Board, <http://www.nhb.gov.sg/~/media/nhb/files/places/trails/jurong/jurong%20heritage_24042015_preview.pdf?la=en> (accessed 14 October 2017).

which side is right but more a matter of following different technical standards. Thus, the Chinese side also took issue with Singapore's high safety measures: Why was it necessary to go for a deep piling of 25 metres for certain tall buildings? Why were the underground sewers buried nine metres deep instead of following the Chinese practice of six metres?

Looking back, all these sounded like technical trivia, unworthy of serious mention for such a large government-to-government (G-to-G) cooperation project as they could be ironed out by people on the ground. However, underlying such different technical standards are the different approaches to development strategies between China and Singapore. China in those days was grappling with sanctions imposed by the West in the wake of the 1989 Tiananmen incident. It was still in its early phases of economic reform, short of capital and lacking the knowledge of the best international practices and technical standards. It had a penchant for what was popularly known as "rolling development strategies", namely, taking short-term projects or even makeshift plans in order to economise on its limited resources for a quick start. The China of today would of course have no problems in embracing any high international standards and in certain areas like its celebrated high-speed train and ubiquitous e-commerce they are in fact blazing a trail. Suffice it to say that Singaporean officials had to stand firm by their high standards, cutting no corners and accepting no trade-offs for a lower requirement. This is precisely the *raison d'etre* for Singapore's involvement in Suzhou in the first place: to build a good industrial park in order to demonstrate Singapore's successful development experiences.

A Highly Uncommon Origin

It was Deng Xiaoping who first provided the political setting for China-Singapore collaboration in Suzhou. During his "southern tour" (*nanxun)* in early 1992, Deng made a distinctive remark to the effect that Singapore had managed to achieve both high economic growth and good social order, and that China should tap on Singapore's experience and "learn to manage better than them". Looking back, George Yeo, then minister for Information and the Arts, once remarked that Singapore was pleasantly surprised when Deng endorsed Singapore as a model of development for China given the country's minute size and population compared to China.

Shortly after Deng's endorsement of the Singapore model, several hundred Chinese delegations from all levels flocked to Singapore to observe and study various aspects of Singapore's economic and social life: public housing, labour market, CPF (Central Provident Fund) system, airport management, development of industrial estates and so on. It soon became clear to Singapore leaders that a country's developmental experience cannot be learnt through a mere visit or by undertaking ad hoc observations. It has to be first systematically packaged before it can be effectively transferred. This gave rise to the Singapore software programme, which was supposed to summarise Singapore's economic management and public administration experience along with the basic institutional structure and operating philosophy behind it.

The transfer of the software programme, however, requires a proper mechanism and a better way to do this is obviously through direct participation in a real project. It is well known that Singapore's town planning expertise was epitomised in its development of the Jurong Industrial Estate. Much of Singapore's public administration experience was best manifested in the way the Changi Airport was built and run. Hence the idea of linking the software programme to an industrial park project, which would enable young officials from both sides to work together for a common development goal. In the process, China could study and learn from Singapore on how to plan, implement and administer an integrated new town comprising industrial, commercial and residential sectors. The idea was enthusiastically embraced by Chinese leaders. Such is the genesis of the SIP.

Not Really "Different Dreams"

However, why Suzhou? The choice of location for the proposed industrial park had not been an easy decision for Singapore to make as the government at that time was also toying with the idea of locating it in Shandong. The location advantages of Suzhou, both then and now, are of course overwhelming in terms of its infrastructure, talent pool, accessibility and prospects for expansion even beyond the initial designated 70 sq km. Every factor weighed heavily in favour of Suzhou except for one thing. Before Singaporeans came, Suzhou had already started its own industrial park in the Suzhou New District (SND). Initially, the small SND looked sufficiently innocuous to Singapore's policymakers who did not fully appraise the important role that it played in the local economy and the degree of local support it enjoyed. More importantly, the local authorities at that time had given Singapore the assurance that it was fully behind the SIP and would prioritise SIP's development over SND.

It soon became clear that the existence of SND posed the single most important obstacle to enlisting whole-hearted cooperation from Suzhou officials for SIP. There were instances where some local officials unofficially promoted SND (regarded as completely their own) to compete unfairly against SIP (regarded as "foreign" or purely Singapore, though in reality SIP was a joint venture of 65-35 between Singapore and China). This gave rise to a great deal of unnecessary conflicts and disputes at the implementation level, culminating in Senior Minister (SM) Lee Kuan Yew's public airing of unfair and debilitating competition in Suzhou in late 1997. In the event, he brought the matter all the way to Beijing to enlist the central government's help to resolve the matter.

From Singapore's point of view, there could only be one team in Suzhou, not two, so that "when we win, they win; when we lose, they lose". With standing conflict between SIP and SND, interests were not aligned and common objectives not shared. Progress was therefore slow. The conflict was subsequently resolved in 1999 after, among many things, an agreement was reached to reverse the original equity structure, with the Singapore consortium lowering its stake to 35% and the Chinese side raising it to 65%.[3] This in turn provided a new incentive structure for Suzhou officials to ensure SIP's success. Since then, both sides have been able to work hand-in-hand in the interest of SIP.

[3] This switch in share ownership was effected in January 2001.

Some media reports have used the description "same bed, different dreams" or the "mountains are high and the emperor is far away" to describe the misalignment of interest between Singapore and Suzhou authorities in the past. However, such a broad description oversimplifies the SIP-SND issue and ignores the intricacies involved on the ground. As former Suzhou Party Secretary Yang Xiaotang once reiterated, the Suzhou government had always accorded priority to SIP because this was a project involving Singapore and China, and its success or failure would have an impact on bilateral relations. However, he also added that as the ranking official in Suzhou, he could not neglect SND because it had an earlier kick-start and was already an integral part of Suzhou's economy in terms of its financial contributions and employment generation. Yang used the term "*shouxin shoubei*" (the palm and the back of one's hand) to justify his equal treatment of both SIP and SND. Other Suzhou officials have also remarked that even if SND was not there, SIP would still have to contend with competition from other neighbouring industrial parks such as Kunshan and Wuxi. To them, this was the harsh reality of market competition.

Hardware and Software

In terms of project design, SIP, in combining the government-to-government software transfer with the commercially oriented industrial park as hardware, was originally considered a very "rational" structure, cleverly conceived. Initially, software transfer was supposed to be the driver of the project while the development of hardware was intended to be a vehicle for software transfer. Theoretically, software transfer and hardware development were separate components and could work in tandem with each other. In this sense, there is a connection between software transfer and hardware development in that the sooner the Chinese side imbibed the principles, values, concepts and practices that the Singapore side sought to share (i.e. its software), the sooner the SIP's physical development (i.e. hardware development) would attain a level comparable to the Singapore experience. However, such ideal design structure did not function well when it was planted on foreign soil. Their inherent incompatibilities soon surfaced.

In the initial years, the Singapore side encountered resistance in imparting its software to the Chinese side. There was a view that the Chinese side did not fully appreciate the value of Singapore's software which was intended to position the SIP in good stead in the long run. As mentioned earlier, the Chinese side could not understand why their Singapore counterparts insisted on pilling to such a depth until serious flooding vindicated this approach. Prior to this, the Chinese side had the impression that the Singapore side was out to make profit at the expense of their Chinese counterparts.

It was also observed that the Chinese side was initially interested only in selective aspects of the Singapore software. For instance, the Chinese side was apparently interested in the kind of fees companies operating in Singapore were levied so that they could charge the same kind of fees on investors that came to SIP. In essence, there was the view that the Chinese side had a tendency to apply Singapore rules and regulations for the purpose of generating revenue for the short term and less on creating an overall conducive environment for SIP investors.

On a few occasions, the two sides had disagreements over the hardware development aspects of the SIP and such differences loomed so large that they overshadowed the significance of the software transfer of the SIP. For example, the joint venture company found itself having to pay for infrastructure-related facilities that were essential to attract investors to the SIP which a commercial entity would normally not undertake. These "public" responsibilities added to the financial burden of the company. There were also differences over Singapore's insistence on adhering to the very letter and spirit of master planning as opposed to the Chinese penchant to make adjustments to any previously drawn up plan so as to take into account exigencies that developed along the way. So instead of the original intention of having software transfer drive the hardware development, much attention was expended by both sides to tackle the differences that arose over the hardware or physical development of the SIP.

Today, with the experience gained over the years, the Suzhou authorities are playing an active role in helping SIP grow and prosper. They are well aware that while the operations in SIP should adhere to strict market principles, the onus is on the local government to create a conducive environment such as securing better terms and conditions, providing better infrastructure and mounting more aggressive and professional marketing campaigns.

Testing Time for Young Singapore Officials

With such strong blessing from high places in both China and Singapore, the Chinese manifestly eager to learn from Singapore and Suzhou Mayor Zhang Xinsheng personally coming to Singapore to lobby for the project to be sited in his city, the young Singaporean officials who were sent to work in Suzhou would naturally expect a plum job. However, they soon realised that they were in for a shock — indeed, quite a lot of "cultural shocks".

It is not that Singapore officials could not speak Mandarin or they did not understand Chinese society. It was more due to problems like different mindsets and working styles, and different bureaucratic cultures. For one thing, the direct problem-solving approach adopted by many young Singapore officials often ruffled the feathers of their Chinese counterparts. Thus, both sides at the implementation level soon got bogged down by arguments and wrangling, as exemplified in the aforesaid cases.

One of the single most important causes for the implementation hurdles was China's complicated multilayer political structure. It is well known that China operates a five-level government and Singapore understood this. Looking back, however, Singapore seemed to have overestimated the ability and willingness of the central government in Beijing to influence events at the lower levels, and underestimated Suzhou government's tremendous influence and control over the SIP project.

From the start, Singapore adopted a top-down approach, on the assumption that as long as Beijing saw eye-to-eye with Singapore on the strategic importance of the project, no major implementation obstacles could arise. How would it still be possible for local officials in Suzhou to drag their feet after President Jiang Zemin had personally endorsed the SIP project as *Zhong-zhong zhi-zhong* ("a top priority of all priorities")? However, things were not so simple on the ground.

China scholars have long realised that the issue of central-local relations in China is extremely complex. It denotes a set of dynamic power relations between the centre and localities, which defies simple characterisation. Suffice it to say that the central government, to safeguard its power and prestige, has always been very selective in intervening in local affairs. Specifically for the SIP project, Beijing did not deem it appropriate for the central government to intervene in all the frictions and disputes concerning SIP which could best be administratively resolved by the local authorities.

Local governments, more often than not, usually have to take into account local peculiarities and conditions when carrying out broad directives or policies issued by the central government. Such a flexibility allows them to engage in what may be termed "selective compliance" given the vast expanse of China and its diversity. Specifically for Suzhou, local officials had all along (until the change in the project's ownership pattern in January 2001) attached greater importance to SND than SIP. To some extent, SIP was regarded as "foreign interests" because of Singapore's heavy involvement. Apart from the need to promote local interests, local officials were fully aware that their career tracks were tied to local bureaucratic loops, not with the central system. Hence many local officials just did not have strong incentives to speed up the SIP process.

As local officials continued to engage in unfair competition against the SIP, the frustrated Singapore officials eventually sought the intervention of SM Lee, who informed central leaders in Beijing that SIP was not proceeding as smoothly and rapidly as expected due to intractable difficulties on the ground. After SM Lee's public airing of Singapore's frustration over its slow progress, the project was subsequently restructured and re-energised, laying the ground for its recent growth.

In recounting the development process of SIP, credit must be given to China's central leadership in Beijing for their unwavering confidence and optimism in the ultimate success of the project, even though they adopted a largely "hands off" attitude towards SIP matters. Central leaders in such a big country as China always tend to look at the "big picture" and take a "long vision". Thus, back in September 1999, when SIP was at its lowest point with an accumulated paper loss of over S$100 million, Vice-Premier Li Lanqing (as China's highest official responsible for SIP, who co-chaired with Deputy Prime Minister Lee Hsien Loong at SIP's Joint Steering Council meeting) remained bullish. He explained to Western media (which were then spreading negative news about SIP): "In the start-up years, it is natural to incur some losses". Then he pointedly told them: "I'm confident the SIP will be profitable some time in the future".

A Miracle in the Making

It took less than two years to prove Li Lanqing was right. SIP today is thriving. SIP as a company started to turn around in 2001 after it had accumulated total losses of US$77 million. In 2001, it started to make small profits of US$7.6 million, and this increased to US$11 million in 2002 and expanded further to US$71 million in 2003. Accordingly, SIP announced its first dividend payout to shareholders during the 10th anniversary celebrations in 2004. It has also been ready for a listing on the Shanghai stock exchange for several years already. This is a far cry from the mistaken image of SIP as a losing concern in the 1990s.

In fact, under the strong efforts of the local authorities and the Suzhou Industrial Park Administrative Committee, SIP has already earned a name for itself and gained recognition as a premier destination for high-tech investments. As early as 2001, Suzhou was already ranked among one of the nine emergent new-tech cities in the world in an article in *Newsweek* while another article in *The New York Times* described Suzhou as transforming from a city of silk to a centre of technology. Elsewhere, Suzhou was described as a "boomtown" (*Wall Street Journal*, November 2001) or even a "new frontier" for companies looking to set up manufacturing bases overseas (*Far Eastern Economic Review*, December 2001). Above all, the per-capita GDP of Suzhou in 2003 (at 47,401 yuan) overtook Shanghai (46,717 yuan) for the first time.

Not surprisingly, SIP has become a magnet for quality FDI (foreign direct investment). By 2017, SIP had more than 5,800 approved foreign-invested projects and actual utilised foreign investment of US$29.4 billion (compared to over 3,000 projects and actual utilised foreign investment of US$16 billion in 2009, the year it celebrated its 15th anniversary). Among the foreign-invested projects, there are so far 92 Fortune-500 companies that have invested in 156 projects in SIP (compared to 77 Fortune-500 companies in 125 projects in 2009).

A never-ending challenge of SIP, like other successful industrial zones and economies elsewhere, is to constantly identify new sources of growth through economic restructuring and industrial upgrading. One of the major transformations in SIP has been the shift from low-end, labour-intensive manufacturing to more technology-intensive or higher value-added activities. Its ambition is to become the "Silicon Valley of the East" by attracting high-tech industries through a slew of measures including discharging relocation payouts, implementing tax breaks, allowing interest-free loans, administering R&D grants, and setting up incubators and creating eco-systems for a series of high-value industries.

Today, the high-tech industrial output accounts for over 66% of the total industrial output. In 2016, the three high-tech sectors of biomedicine, nanotechnology and cloud computing generated total sales of 47 billion yuan, 38 billion yuan and 35 billion yuan, growing by 57%, 36% and 25% respectively from 2015. SIP remains an attractive location for high-tech companies to either set up shop or for expansion purposes. In 2017, apart from existing big players like Huawei, Microsoft, Siemens and Roche Diagnosis, Apple announced that it will establish a research centre in SIP.

Apart from the industrial sector, the commercial and residential sectors are also experiencing rapid growth and development. Not only is SIP conducive for doing business but it is also known for offering a quality living environment, thanks to the Singapore-style urban planning and management practices, and concerted greening efforts. It is worth highlighting that Suzhou, with SIP being an integral part of it, was conferred the Lee Kuan Yew World City Prize in 2014.[4] The SIP has also successfully tackled the problem of urban villages or urban sprawl prevalent in many other cities in China by integrating urban villages into a unified urban administrative network. In other words, farmers who were previously living in SIP now have access to urban amenities like their city counterparts, thereby promoting greater urban equality and by extension, social harmony.

On the software side, SIP continues to pride itself as being highly "business-oriented", offering a "one-stop" shop to investors. Going all-out to please the investor is the *modus operandi* of SIP. Speed, efficiency and professionalism in meeting customers' needs have become the norm. So far, over 3,000 Chinese officials have come to Singapore for training in over 160 different courses since 1994. Of equal significance, several hundreds of thousands of officials from different parts of China have visited the industrial park to observe and learn from its successful experience. In a sign that software transfer has become mutually beneficial, Enterprise Singapore (formerly known as IE Singapore) launched a 12-day China Ready Programme in 2017 for business executives to gain first-hand knowledge of China's business environment such as market insights, culture and financial systems in Suzhou. Among the highlights of this programme is the opportunity to interact with representatives from companies in not only SIP but also Suzhou as a whole.

SIP has gone on to set up mini-SIPs in other parts of Jiangsu and other Chinese provinces like Xinjiang and Anhui. In doing so, the local authorities have stressed that SIP is "going out", building on the successful experience of SIP. Inherent in these remarks is a recognition that SIP is different from other industrial parks because of the Singapore

---

[4] The Lee Kuan Yew World City Prize is a biennial international award that honours outstanding achievements and contributions to the creation of liveable, vibrant and sustainable urban communities around the world. For more details, please refer to <https://www.leekuanyewworldcityprize.com.sg/about_prize.htm> (accessed 24 October 2017).

connection or Singapore DNA. In this sense, the success of SIP has enhanced the "Singapore brand name" in China and continues to be a boon for Singapore enterprises either operating or thinking of venturing into the vast Chinese market.

No Mean Feat

SIP owes a great deal of its success to the strong commitment and unstinting support of the two governments. From the start, the two governments were determined to build a high-quality industrial park that would set the standard for both existing and future industrial parks in China, in terms of not only good infrastructure and facilities available but also how it would be run and managed. Their determination and faith had never wavered even when the project had run into implementation hitches.

From Singapore's view, it had sent its "best and brightest" there and they had spent a considerable amount of time and effort to market SIP internationally. Singapore's top leaders often made it a point to promote the project to foreign governments during their overseas tours. China on its part had also fully supported the project by granting favourable policies and other facilities. Apart from Li Lanqing who directly oversaw the project, other central leaders such as Jiang Zemin, Li Peng, Zhu Rongji and Qiao Shi had personally visited SIP to lend their support.

However, as mentioned earlier, Beijing's strong support for this project does not automatically translate into a "hands-on" policy for SIP where central leaders would break bureaucratic convention to directly intervene in how the Suzhou government should deal with SIP matters. From the start, the Singapore side stressed that the industrial park had to proceed in accordance with the market principle. China's central leaders soon came to appreciate this basic tenet. Ironically, if China's central leaders were to behave much like old-time central planners of a command economy whereby detailed administrative guidelines would be issued to the Suzhou government on how to proceed with the project, a lot of implementation hitches might never have cropped up. However, this is exactly something the two governments would not want.

Looking Back and Moving Forward

As it happened, the implementation of the project in terms of its physical development was fraught with squabbles and wrangling, particularly in the early phases. A great deal of unnecessary tussles stemmed from the fortuitous presence of the rival park, SND, a major contributor to the Suzhou economy. However, the differences in respect of Suzhou and Singapore officials' mindset, bureaucratic cultures, working styles and expectations had also aggravated the situation.

Looking back, it is fair to say that quite a lot of the original implementation tussles were much in the nature of such a large and complicated project, especially when China in those days had just opened its door wider in response to the calling of Deng Xiaoping. Its institutional framework was still weak and it lacked experiences when dealing with foreign businesses. Even if SIP were to start off as a completely private concern run by purely commercial groups, it would still have experienced a lot of these hurdles.

Many large multinationals, which went to China in the early 1990s, had to similarly cross a lot of hurdles before they finally started to reap profits in the recent years. General Motors and its experience with the Beijing Jeep were a case in point. They learned from their mistakes and then thrived.

Putting past difficulties in a positive light, the SIP has helped both sides to get to know each other better and faster as SIP is not purely a commercial project. At the political level, the SIP allows leaders and officials on both sides to interact at numerous joint council and steering council meetings as well as at many other informal and formal occasions. The official and more importantly, personal and business networks that were built up over the years would continue to be valuable in the deepening and broadening of future China-Singapore relations. It may even be surmised that such a level of interaction through the various councils is likely to become even more significant when a younger and different generation of leaders take the helm in both Singapore and China. In this sense, there are plenty of positive spillovers from SIP that are intangible and unquantifiable particularly when viewed from a longer-term perspective.

Outline of Chapters

This volume arose out of the authors' continuing interest in examining how SIP was faring in the midst of China's national push for further reform and opening up in general and the Suzhou authorities' emphasis on economic upgrading and restructuring in SIP in particular. As an industrial park that faces intense competition from other industrial parks and development zones across China, we wanted to take stock of the progress and challenges faced by SIP as it sought to innovate and move further up the value chain. We expect this process to be incremental (and not drastic) in nature and more difficult to achieve given that SIP is progressing from an already high stand-point.

Apart from taking stock of SIP's economic achievements and challenges, we are also interested in SIP's progress on the social development front. Since its inception, the SIP was designed not only as an industrial park, but also as a conducive place for residents to live, work and play with the commercial and residential components included. Efforts to create such a liveable environment might not have been obvious in the initial years of the project as the focus then was on bringing investors into the SIP. However, over time, the key elements of liveability such as having a clean and green environment, convenient access to daily amenities and facilities, availability of good schools and efficient transportation networks became more pronounced and a visible selling point for the industrial park. We have therefore also wanted to take stock of SIP's social development and, more importantly, how progress in this area has helped to enhance the attractiveness of SIP not only for investors but also for their family members and the highly skilled talent required to raise SIP's development to a higher level.

In fact, we are beginning to feel that SIP's social development in the long run may even be more significant and lasting, especially from

Singapore's point of view. SIP's economic and industrial achievements are no longer unique in China as many cities in China today such as Shenzhen have also achieved or even surpassed SIP's levels. However, many salient aspects of SIP's social development from urban planning to public housing to social management at the neighbhourhood level have remained unique among China's cities with their strong Singapore characteristics. With China under Xi Jinping currently emphasising the need for better-quality growth, SIP's remarkable progress in those social areas could be a showcase of its pioneering role in China's socio-economic development. In retrospect, Singapore leaders at the beginning had probably not expected that some areas of SIP's social development had continued to embody certain Singapore's distinctive legacies.

To provide readers with some useful background of the SIP project, Chapter 1 gives a brief historical overview of the project, tracing how the SIP started and highlighting some key lessons of experiences when the Singapore side was the majority owner and shareholder of the project from 1994 to 2001. It also traces the origins of the high-level bilateral cooperative mechanism between Singapore and China in existence today and the bilateral cooperative mechanism established to drive the SIP in its initial years. In the same vein of providing relevant context, Chapter 2 highlights the level of importance the Chinese side has continued to accord to the SIP especially after it has taken over as the majority owner and driver of the project since 2001 and gives an account of the broad economic and social changes that have occurred in the SIP over the years.

As mentioned earlier, a key thrust of this publication is to examine how far SIP has progressed on the economic front. Chapter 3 outlines the different development phases of the SIP that essentially traces its transformation from low-end, labour-intensive manufacturing to more technology-intensive or higher value-added activities. To be sure, SIP is facing factory closures and relocation as less competitive industries are being phased out. Despite this development, SIP has been able to attract innovative and higher value-added industries that more than make up for the loss in production value. Chapter 4 highlights the three high-tech innovative industries of biopharmaceuticals, nanotechnology and cloud computing that SIP seeks to nurture and grow. Unlike the traditional export-oriented industries, these three high-tech sectors are geared towards the domestic market and rely on high-skilled labour. In this sense, they tend to be less susceptible to the "foot-loose" behaviour of companies when labour and business costs rise.

Another key focus of this publication is to highlight the progress of SIP on the social development front. Chapter 5 outlines the distinctive features of SIP's social development which includes the development-oriented nature of its social policies, the fine balance it has managed to strike between creating a liveable environment and pursuing development, and its success in integrating what was previously urban villages into an integrated, urban administrative framework. Chapter 6 dwells on the social foundations of SIP's industrial upgrading which is underpinned by its ability to attract top talents and high-tech set-ups. It does so by providing good social amenities and environment which based on our ground interviews is the most important factor apart from the various schemes and incentives that the local administrative authorities have introduced to attract the requisite talents. An obvious fact is that SIP's earlier investment in social development is now reaping dividends. More than simply being the outcome of a carefully designed plan, social development has now become a key source of SIP's "comprehensive competitiveness" that makes it stand out from other industrial parks and development zones across China.

The East Asian Institute (EAI), as Singapore's institute doing empirical and policy relevant research on China, has all along been paying close attention to SIP's growth and development. Way back in 2001, we were involved in a research project commissioned by Singapore's Ministry of Trade and Industry, to highlight the lessons of experience when the Singapore side was the majority owner and shareholder of the project from 1994-2001.[5] Based on extensive field interviews conducted in Singapore and China, EAI submitted a detailed report to the

[5] Apart from Prof John Wong and Mr Lye Liang Fook from the East Asian Institute (EAI), the other individual who was involved in this detailed study was Prof Zheng Yongnian who was then a senior research fellow of EAI.

Singapore government. The report remains unpublished as it is owned by the Singapore government.

Since then, we have closely followed SIP's development and submitted occasional reports to the Singapore government based on short-term study trips to SIP. This publication, that highlights some of the latest developments in SIP, is based on our recent study trips and we are happy to reproduce here for the interest of general readers. A primary motivation behind this publication is our sense that the general public perception of SIP in Singapore has tended to remain stuck in the earlier years when the Singapore and Chinese sides encountered intractable difficulties in developing the project. While not forgetting what has transpired in the past, we feel that it is important to situate this project in the proper context. More specifically, we wish to generate greater public awareness of how far this project has moved way beyond its initial years to one where it is today recognised as a leading industrial park in China, not only in terms of its economic achievements but also for its pleasant and liveable environment. By continuing to work together, the SIP offers a regular platform for the governments of the two countries to deliver further progress on the project, which will in turn add substance to the burgeoning bilateral relationship.

We wish to acknowledge here that our study trips have been made possible due to the strong support and generous interviews granted by the leadership and management of the Suzhou Industrial Park Administrative Committee, the local authority directly responsible for overseeing SIP's development. We also wish to thank the China-Singapore Suzhou Industrial Park Development Group Pte Ltd, the joint venture company that spearheads the development of the China-Singapore cooperation zone within the SIP, for their unstinting time and effort in helping us gain a more rounded perspective of the interests of companies operating in SIP.

The authors further wish to acknowledge that this publication is the product of a research team of five EAI scholars led by Prof John Wong and comprising Mr Lye Liang Fook, Dr Zhao Litao, Dr Henry Chan and Ms Ping Xiaojuan. Finally, we hope that readers will find this a useful read as much as we have enjoyed in coming up with this publication.

Chapter 1

No Ordinary Industrial Park

Introduction

The Suzhou Industrial Park (SIP) that began in 1994 is Singapore's first flagship G-to-G project in China. This chapter provides a background to the origins of SIP, highlights some lessons of experiences when the Singapore side was the majority shareholder and outlines the high-level bilateral cooperation mechanism that has evolved over time. The origins of SIP could be traced to Deng Xiaoping's 1978 visit when he was struck by Singapore's socioeconomic transformation under Prime Minister Lee Kuan Yew. Deng further cited Singapore as a reference model during his Southern Tour in 1992. At that time, Singapore was on an active regionalisation drive and Lee Kuan Yew conceived the idea of the two countries working jointly on a project in the form of the SIP to impart Singapore's experience in economic management and public administration to Suzhou.

During the initial years, the Singapore side as the majority shareholder and manager of the project encountered a number of challenges. Some of the key challenges involved inevitable start-up problems, while others were the clash of personalities with the attendant mindset and personal networks; the different bureaucratic cultures, value systems and approaches to problem solving; and the Chinese side's resistance to the perceived "over Singaporisation" of the project. Other difficulties encountered included the asymmetrical decision-making process on both sides; the different priorities accorded by the Singapore and

Suzhou authorities to SIP's development; the original business model of SIP that worked against its commercial viability; and operational issues relating to having software development underpin the hardware or physical development of the project. Many of these challenges were subsequently addressed or resolved in agreements signed by the Singapore and Chinese sides in 1999, which included handing over majority ownership and management of the project to the Chinese side from 2001.

To drive SIP's progress, Singapore saw the need to institute a high-level bilateral mechanism that enlisted Beijing's support from the very beginning. In fact, this deputy prime minister-led mechanism was proposed by Singapore as early as 1993, a year before the SIP started. Such a mechanism has expanded and been augmented over the years. In 2003, Singapore and China launched the Joint Council for Bilateral Cooperation (JCBC) that oversees the development of SIP. Today, the JCBC is also responsible for two other G-to-G projects, namely, the Tianjin Eco-city and Chongqing Connectivity Initiative. Operationally, the JCBC helps to drive each of the three G-to-G projects. It also serves a higher function of offering a platform for the younger generation of leaders from both sides to get to know each other better. This has been a prevailing objective that has perhaps become even more important in the post-Lee Kuan Yew era.

A Highly Unusual Origin

The origin of SIP could be traced as far back as November 1978 when Chinese leader Deng Xiaoping visited Singapore in one of his rare overseas visits.[1] Deng was particularly struck by Singapore's socioeconomic transformation that was starkly different from official Chinese stereotypical accounts of Singapore and the old impression he had of Singapore as a backward entrepot during his last transit in 1920.[2]

[1] Besides Singapore, Deng visited Thailand and Malaysia in November 1978.

[2] In 1920, Deng spent two days in Singapore while en route to France for a work-study programme after the end World War I. Deng was around 15 years old at that time.

Deng had since closely followed developments in Singapore. Fourteen years thereafter, in early 1992 during his *Nanxun* (Southern Tour) to revive the momentum of China's open door and reform policy stalled by the 1989 Tiananmen incident, Deng specifically singled out Singapore as a reference model for China's development in his *Nanxun* speech.

More specifically, Deng said that "Singapore apart from achieving rapid economic growth also enjoys good social order. They govern the place with discipline. We should adapt their experience and learn how to manage better than them".[3]

Deng's specific mention of Singapore provided the political imprimatur for Singapore and China to cooperate further. It sparked off a "Singapore fever" in China, leading to the visit of more than 400 Chinese delegations to Singapore in 1992 alone to better understand how Singapore had established good social order alongside rapid economic growth.[4] There were however limits to how much these delegations could learn from Singapore given the short duration of their observation tours.

In September 1992, when Lee Kuan Yew (then Singapore's senior minister) visited Suzhou, its Mayor Zhang Xinsheng broached the idea of Singapore investing a part of its reserves to help Suzhou industrialise. Lee was however not immediately seized with the idea of collaborating with Suzhou as it was unclear to him whether and to what extent China's central government would be supportive of Zhang's proposal.[5] Only after several rounds of meetings and discussions was the idea of Singapore developing an industrial park in Suzhou conceived.

On its part, Singapore had around the early 1990s pushed for Singapore companies to go regional to develop an "external wing".[6]

---

[3] Shenzhen Propaganda Department, ed., Deng Xiaoping yu Shenzhen: 1992 Chun (Deng Xiaoping and Shenzhen: Spring 1992), Shenzhen, Haitian chubanshe, 1992, p. 9.

[4] John Wong, "China's Fascination with the Development of Singapore", *Asia-Pacific Review*, vol. 5, no. 3, Fall/Winter 1998, pp. 51–63.

[5] Lee Kuan Yew, *From Third World to First: The Singapore Story, 1965–2000*, Singapore, Singapore Press Holdings, 2000, p. 719.

[6] "Singapore's Second Wing Looks Set to Take Flight", *The Straits Times*, 17 April 1994; "Ventures Abroad: Panel of Advisers Named", *The Straits Times*, 31 January 1993; and

Specifically, at a local enterprise conference in 1992, Deputy Prime Minister and Trade and Industry Minister Lee Hsien Loong said that due to the "physical limits of growth in Singapore, especially manpower", Singapore companies had to "venture overseas, to take advantage of opportunities elsewhere and operate in the international marketplace".[7]

Likewise, in his address to People's Action Party cadres in 1992, Senior Minister Lee Kuan Yew cited the success of the outward-looking development strategies of Taiwan, Hong Kong and South Korea, highlighting that these three newly industrialising economies had "two wings with which to take flight. With only one wing, Singapore will stay on the ground and not get airborne".[8]

In its "going out" strategy, one of Singapore's key approaches was for the government to select a suitable province or state with the "potential to be an NIE" and whose leaders are keen to tap on Singapore's "experience and expertise in drawing up and executing their economic development plans". This approach would enable Singapore to build a "broad and deep relationship" with the local leaders, which in turn could be helpful to "Singapore businessmen in securing some of their development projects".[9] Such was Singapore's approach to developing economic ties with Johor, Malacca and the Riau province at that time. It was broadly the same approach for the proposed SIP project.

The SIP was conceived due to such a fortuitous confluence of factors on the part of Singapore and China. From China's perspective,

---

"SM Lee: Singapore Will Become Failed NIE If Its People Do Not Venture Abroad", *The Straits Times*, 3 January 1993.

[7] "Keynote Address by BG Lee Hsien Loong, Deputy Prime Minister and Minister for Trade and Industry, at the Enterprise '92 Conference", National Archives of Singapore, 25 August 1992, <http://www.nas.gov.sg/archivesonline/data/pdfdoc/lhl19920825s.pdf> (accessed 14 February 2017).

[8] "SM: Singaporeans Must Now Build Up External Economy", *The Straits Times*, 16 November 1992.

[9] "Keynote Address by Prime Minister Goh Chok Tong at the Regionalisation Forum", National Archives of Singapore, 21 May 1993, <http://www.nas.gov.sg/archivesonline/data/pdfdoc/ ct19930521.pdf> (accessed 31 March 2015).

Singapore was a useful reference for its economic development. Singapore had struck a positive image in Deng Xiaoping's mind since 1978 and Deng further gave it a ringing endorsement in 1992.

Singapore was also politically more acceptable to China than Hong Kong (which was still under British rule at the time), Taiwan (which was regarded as a province of China) and South Korea (which did not establish diplomatic relations with China until August 1992).

From Singapore's perspective, Deng Xiaoping's special mention of Singapore in his 1992 speech, albeit unexpected, dovetailed well with its regionalisation strategy. The SIP was thus conceived to enable both sides to work "hands-on" for a common pursuit, which was to transfer Singapore's experience in economic management and public administration to Suzhou. Software transfer was intended to be the driver to underpin SIP's overall development.

Some Lessons of Experiences

To develop the SIP project, a joint venture company known as CSSD was formed comprising companies from a Singapore consortium and a Chinese consortium.[10] A joint venture company would underscore the importance of developing the SIP on a commercial basis. While the two governments would collaborate closely on software transfer, the SIP project had to be commercially viable to be sustainable.

For almost seven years from the start of the SIP project in 1994 to 2001, the Singapore consortium held 65% ownership and management of the project while the Chinese consortium held the remaining 35%. A number of public sources have highlighted the difficulties

---

[10] CSSD is the joint venture company responsible for spearheading the development of the initial 70 sq km Sino-Singaporean cooperation zone within the SIP (which was subsequently expanded to 80 sq km). It was initially known as the China-Singapore Suzhou Industrial Park Development Co Ltd (or CSSD). It was renamed the China-Singapore Suzhou Industrial Park Development Group Co Ltd (with the same acronym CSSD) in June 2009 as a key step in the IPO process. From the CSSD side, the Singaporeans who have helped the study team with its interview arrangements were Mr Bernard Teo, Mr Goh Kian Seng and Ms Cheong Yiling.

Singapore had encountered in implementing this project during this period. The difficulties can be broadly divided into systemic and incidental factors.

Systemic factors refer to those that originate from or are related to the existing political/institutional structure or systems. Incidental factors are more like random variables that are associated with particular individuals or personalities with their attendant mindsets and personal networks. In theory, it is relatively easy to separate the systemic or incidental factors, but in reality, these factors are intertwined and not easy to compartmentalise.

A key incidental factor that affected cooperation was the role of different personalities and their attendant mindset and political/bureaucratic networks. In particular, Suzhou Mayor Zhang Xinsheng who was responsible for the project at the beginning was extremely protocol and status conscious. As mentioned earlier, Zhang had suggested to Lee Kuan Yew to set aside a part of Singapore's reserves to help Suzhou industrialise.[11]

To build bridges with the Chinese side and get the project off the ground, the first Singapore CSSD chief executive officer (CEO) was selected for his fluency in Mandarin and familiarity with Chinese classics.[12] Zhang, however, regarded the first Singapore CEO as a political lightweight. He felt that Singapore should have sent someone more senior and with more standing, preferably someone who was linked to the EDB (Economic Development Board), the premier institution responsible for bringing investments into Singapore.[13]

---

[11] Zhang Xinsheng was Suzhou mayor from 1989 to 1997.

[12] The first CEO of CSSD's term of office was from 1994 to 1996.

[13] Another complaint that Zhang had was that the first CEO of the CSSD was not based full-time in Suzhou. Zhang felt that this was necessary as investors would like to view the SIP site and it was therefore important for the CEO of CSSD to do on-site marketing. However, according to Chan Soo San, he had spent one-third of his time in Suzhou, one-third in Singapore and the final third engaging in international marketing for SIP. Even for the time spent in Singapore, he was busy doing liaison work, financing and marketing for SIP. In his view, off-site marketing for SIP was equally, if not, more

When an EDB team eventually took over the helm of the CSSD in 1996, a new set of challenges surfaced. While EDB succeeded in bringing more international investors to the SIP, its participation distinctly shifted the working culture and approach more in line with the Singapore practice. A key focus of the second Singapore CEO (who was previously from EDB) was to return CSSD to the path of commercial viability. To do this, certain structures and procedures had to be in place, and issues and problems that had impeded CSSD's commercial viability had to be identified and tackled head-on. Such a business-like and sort off no-nonsense approach ran up against the traditional Chinese penchant for an ambiguous, informal and even face-saving approach.[14]

The role of personalities had also impeded cooperation among various bureaucratic interests groups in Suzhou. Specifically, Mayor Zhang Xinsheng was regarded as an "outsider" by Suzhou officials as he was sent by Beijing for a stint in Suzhou.[15] As mayor and concurrent

important for SIP. It therefore made no sense for him to take up an apartment in Suzhou when it would be empty two-thirds of the time.

[14] For example, the second Singapore CEO took issue with SIPAC's unilateral move to take over and even develop two choice plots of land next to the scenic Jinji Lake (金鸡湖) without formal approval. According to the CEO, what SIPAC did went against the very letter and spirit of master planning, a hallmark of Singapore's software. Despite CSSD's objections, SIPAC went ahead to build, *inter alia*, a government guesthouse, a Chinese euphemism for a hotel. When the guesthouse was completed, the Singapore side refused to supply water to the guesthouse, showing its displeasure with what SIPAC did. Nor did they patronise the guesthouse. This issue was eventually settled as a package in a commercial agreement reached between the Singapore and Chinese sides in June 1999. The Chinese side eventually paid for the two plots of land they took and the master plan for that area had to be amended to take into account the hotel and other structures that were built.

[15] Before becoming Suzhou mayor, Zhang was vice chairman of China's National Tourism Administration from 1986 to 1989 in Beijing. He studied in America from 1980 to 1981. From 1982 to 1985, he was the deputy director of the Tourism Bureau of the Jiangsu provincial government. In the late 1970s to early 1980s, he was the co-founder and executive managing director of Nanjing Jinling Corporation whose flagship project was the Jinling Hotel (金陵饭店) in Nanjing, one of China's first five-star hotels. After he left SIP

chairman of SIPAC (Suzhou Industrial Park Administrative Committee), he guarded SIP as his "baby" and prevented other local Suzhou officials from claiming credit for SIP's development.

Zhang's cosmopolitan outlook and good command of English was well-received, especially by foreigners, who tended to regard him as an effective "marketing person" for SIP. However, back home, these same qualities alienated him from his local Chinese colleagues. They regarded him as being too Westernised and even articulating Western interests. His habit of showing off his command of English in their presence further irked them.[16]

An important systemic factor that affected cooperation between the Singapore and Chinese sides was the asymmetrical decision-making process. In Singapore, a decision made at the top is promptly carried out by the rank-and-file. In contrast, under China's multi-tiered administrative structure, a decision made at the central level usually takes time to trickle down and even then, the lower levels often have some leeway to factor in certain local conditions during the implementation process. Moreover, each administrative level has certain designated roles and responsibilities — the central level is concerned more with setting directions and policies while the lower levels are preoccupied with implementation details.

Given its more complicated administrative structure, the Chinese side was unable to keep pace with Singapore's expectation for efficient decision-making and quick results. At times, when matters could not be resolved at the Suzhou level, the Singapore side would raise the matter one level up to the Nanjing level or if necessary to Beijing. To the Chinese side, this direct and straight-forward approach was done to deliberately apply pressure or to highlight the incompetence of the lower levels.[17]

---

in 1997, he went to Harvard for a master's degree before returning to China to become vice minister of education.

[16] Moreover, Zhang Xinsheng was "often late at coming to meetings". Zhang became increasingly ineffective and was eventually sidelined.

[17] One local official recounted to the study team: "This was not what a friend would do to another friend".

Another systemic factor was the different priorities accorded by Singapore and Suzhou authorities to SIP's development. From Singapore's perspective, since SIP was a flagship project between the two governments, the local authorities ought to give its whole-hearted support. In fact, the local authorities had assured Singapore that it would accord priority to SIP over the Suzhou New District (SND) which was established earlier. This was easier said than done.

Suzhou Party Secretary Yang Xiaotang justified that he had to consider not just SIP's interests alone, but also Suzhou's overall development.[18] Yang once vividly likened the symbiotic relationship between SIP and SND as the "flesh on both his palm and the back of his hand" (*shou xin shou bei dou shi rou*). In other words, both SIP and SND were equally important from his perspective.

Such an approach was naturally unacceptable to the Singapore side because Suzhou from the start was committed to provide full support to SIP. In fact, Yang was not always even-handed towards both SIP and SND. At times, the Singapore side observed that he even promoted SND at SIP's expense.[19]

SIP was further squeezed commercially by the fact that the original business model of SIP worked against the commercial interests of CSSD. As the lead developer, CSSD took on unnecessary and additional financial burdens which a commercial company would not normally do. It had to buy land (both saleable and non-saleable) from the local government before the land could be sold to potential

[18] Yang Xiaotang was Suzhou party secretary from 1994 to 1998.

[19] The competition between SIP and SND gained prominence during the visit of a Keidanren delegation to Suzhou in April 1996. When the delegation visited SIP, they were received by Mayor Zhang Xinsheng. However, when the delegation was brought to SND after visiting SIP, it was received by the more senior Yang Xiaotang who even hosted lunch for the delegation. This difference in treatment gave the Japanese delegation the impression that local priority was on SND and not the SIP.

Yang Xiaotang's absence in SIP was immediately reported by the Singapore side to the Chinese authorities higher up. Yang later received a phone call from Governor Zheng Silin in Nanjing enquiring why he had hosted lunch for the Keidanren delegation in SND. Yang however felt that this was merely a small matter and that the Singapore side had read too much into it.

investors. The land price was based on a pre-agreed formula that escalated on an annual basis. CSSD was also required to provide public infrastructure and some social and public amenities. All these commitments bogged down CSSD financially.[20]

Yet another systemic factor had to do with the project design. The original design was for software transfer to be the driver of the SIP project while hardware or physical development was meant to be the carrier for software transfer. In other words, the hardware development of the project would depend on how fast or how well the Chinese side could imbibe and implement Singapore's experience in economic management and public administration. The symbiotic relationship between software transfer and hardware development was initially considered a very "rational" structure, cleverly conceived.

However, in practice, instead of software transfer akin to being the horse that led the cart, disputes related to the physical development of SIP loomed so large that they overshadowed software transfer. At that time, for instance, the Chinese side could not fully grasp the value of integrated development,[21] objected to Singapore's thorough but much more expensive way of doing things (such as its more stringent requirements for landfill, piling and laying of underground sewers, and high management fee) and took issue with its marginal role in marketing the SIP. They also had difficulty grappling with the open and transparent way of tendering for projects.

To drive the project to meet certain standards and in view of the need to deliver results, the Singapore side needed to have a majority say over decision-making and operational aspects of the project. Unwittingly, this led to what the Chinese side perceived as "over

[20] For instance, though Suzhou Mayor Zhang Xinsheng had promised to provide power to SIP, the local power bureau argued that Zhang did not have the authority over its operations. In the interest of time, CSSD entered into an agreement with Keppel Engineering to build and operate a diesel power plant in the start-up area of SIP. CSSD also had to fork out money to build an exit at Weiting to connect SIP to the Shanghai-Nanjing Expressway. All these activities added to CSSD's financial woes.

[21] Singapore adopted a long-term integrated approach based on master planning while the Chinese side's penchant then was for rolling development, a piecemeal and ad hoc strategy of developing a piece of land whenever a demand for it arises.

Singaporisation" of the whole project, which caused a great deal of unhappiness as the Chinese side felt sidelined and neglected. They thus did not have a strong incentive to fully back the SIP.

Suffice it to say that the Singapore side faced many challenges at the start-up phase of this project — some of these are in fact unavoidable for such a big G-to-G project. China in those days had just reopened its door wider in response to Deng Xiaoping's call. Its institutional framework was still weak and it lacked experience and even credibility when dealing with foreign businesses. China's international image was severely dented by the 1989 Tiananmen incident that saw Western countries imposing sanctions on China. That was a primary reason why the Singapore side held majority share and management of the project in the initial years. Even if SIP were a completely private concern run entirely by commercial groups with governmental involvement, it would likely have to grapple with these hurdles.

Many of these challenges were subsequently either addressed or quickly resolved following agreements signed by the Singapore and Chinese sides in 1999. Thereafter, a key priority of the third Singapore CEO was to prepare for the handover of the SIP project to the Chinese side, a process formally completed in January 2001.[22]

Institutionalisation of Bilateral Cooperation

The political significance of the SIP has often been overlooked as the public's perception of this project has tended to remain fixated on the difficulties Singapore faced when it was the majority owner and manager of the project. From the very beginning, Singapore saw the importance and long-term benefits of having an institutional framework that involved Beijing to ensure uniformity of policy and singleness of purpose at the lower levels.

---

[22] In January 2001, the ownership structure was switched with the Chinese side owning 65% while the Singapore side held 35%. In August 2005, CSSD admitted three new shareholders, namely, Hong Kong and China Gas Investment Ltd, CPG Corporation and Suzhou New District Hi-Tech Industrial Company. This led to a redistribution of ownership of 52% for the Chinese consortium, 28% for the Singapore consortium and 20% for the new shareholders.

This high-level institutional framework served an additional purpose of providing a platform for top leaders and officials from both sides to meet regularly and work jointly on the SIP. Besides ensuring the progress of SIP, the framework helped to promote personal interactions, build long-term relationships and strengthen bilateral ties. This political function remains very much at play over the years and has now evolved into an augmented framework in the form of the JCBC, chaired by a deputy prime minister from both sides.[23]

In his letter to Vice Premier Zhu Rongji in May 1993, months before the project officially started, Lee Kuan Yew proposed a mechanism to drive SIP's development (Figure 1). At the top was a Joint Ministerial Council, co-chaired by then Deputy Prime Minister Lee Hsien Loong and Vice Premier Li Lanqing. Meeting at least once a year, the council would chart the overall direction and scope of the project, approve necessary resources and review progress periodically. It brought together various central ministries and agencies from both sides to facilitate SIP's development.

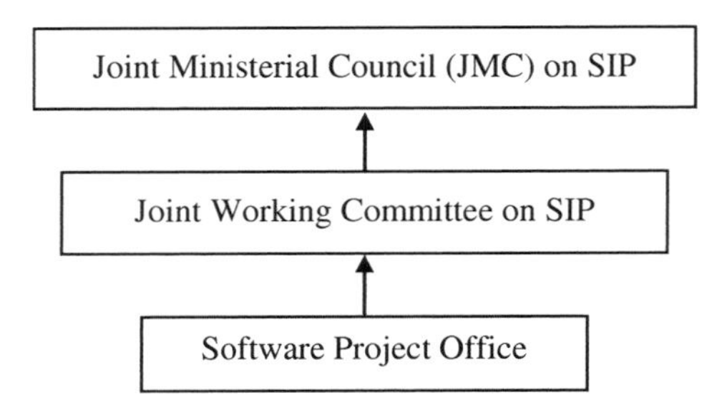

Figure 1. Initial Cooperation Platform

Source: Drawn up by author.

[23] During the recently concluded JCBC meeting in Beijing in February 2017, Singapore's Deputy Prime Minister Teo Chee Hean reportedly said that he had included younger ministers in his delegation so that they can "continue the relationship that was built by senior leaders, Mr. Deng Xiaoping and Mr. Lee Kuan Yew, through the generations". Such interaction among the younger generation of leaders on both sides was a key consideration from the very beginning. See "Singapore Building Up Next Generation of Ties with China", *The Straits Times*, 28 February 2017.

The Joint Working Committee would translate the overall direction and scope of the SIP project as decided by the JMC into specific plans and supervise their implementation. It would meet more often and report directly to the JMC. On each side was a project office responsible for identifying and coordinating areas of software transfer. The software transfer between Singapore and China has continued today.

The SIP institutional framework has been upgraded and expanded over the years. In November 2003, Prime Minister Goh Chok Tong and Premier Wen Jiabao launched the JCBC, the highest-level governmental body between the two countries to promote and facilitate bilateral cooperation. The JCBC was initially co-chaired by Deputy Prime Minister Lee Hsien Loong and Vice Premier Wu Yi on the Chinese side.[24] Today, they are co-chaired by Deputy Prime Minister Teo Chee Hean and Executive Vice Premier Zhang Gaoli.

Under JCBC's purview are three Joint Steering Councils or JSCs (Figure 2): one (renamed from the Joint Ministerial Council) oversees the SIP (mentioned earlier); the second overlooks the development of the Sino-Singapore Tianjin Eco-city (the second G-to-G project); and the most recent is responsible for the Chongqing Connectivity Initiative (the third G-to-G project).

The co-chairs of the JSCs are also the co-chairs of the JCBC. In terms of their function, the JSCs for the Eco-city and Chongqing Connectivity Initiative, like the JSC for the SIP, bring together relevant government ministries and agencies from both countries to ensure a coordinated and comprehensive development of the respective projects.

Under the JSC for the Eco-city, there is a Joint Working Committee that is co-chaired by Singapore's Minister for National Development and China's Minister of Housing and Urban-Rural Development. The Joint Working Committee for the Chongqing Connectivity Initiative is co-chaired by Singapore's minister in the Prime Minister's Office and China's minister of commerce.

---

[24] The Singapore co-chairs of the JCBC were Deputy Prime Minister (DPM) Lee Hsien Loong, DPM Wong Kan Seng and DPM Teo Chee Hean (current). The China co-chairs were Vice Premiers Wu Yi and Wang Qishan and Executive Vice Premier Zhang Gaoli (current).

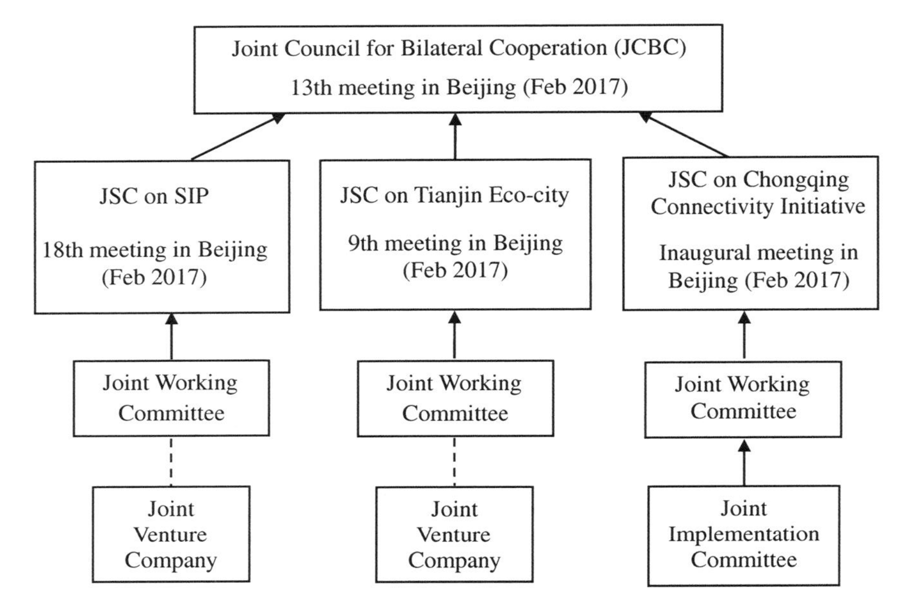

Figure 2. Current Cooperation Platform

Source: Drawn up by author.

What makes the Chongqing Connectivity Initiative stand out from the two earlier G-to-G projects is an additional body in the form of a Joint Implementation Committee, co-chaired by Singapore's minister in the Prime Minister's Office and mayor of Chongqing.

Having a Joint Implementation Committee below the Joint Working Committee is meant to better coordinate policies and implementation at both the national and local levels in China. The Joint Implementation Committee also serves somewhat as a lead "developer" or "driver" of the Chongqing Connectivity Initiative in the absence of a joint venture company or master developer (which is present in the SIP and Tianjin Eco-city projects).[25]

[25] Strictly speaking, the joint venture companies formed for the SIP and Eco-city are not part of the official cooperation platform. They do not officially report to the JWC or to the JSC though there would be informal channels of communication between the company representatives on the one hand and government leaders and officials on the other hand. The relationship between the two is thus denoted by a dotted line in Figure 2.

Under the current JCBC structure, there is an impetus or imperative for each of the three G-to-G projects to continue to make progress and remain successful. By doing so, they will lend substance to bilateral relations.

The JCBC mechanism further offers a valuable platform for both sides to regularly interact and get to know each other better. For Singapore in particular, the mechanism and the projects that underpin such a mechanism offer a means to engage China whose attention is invariably more drawn to its ties with other major powers. It offers Singapore an institutional means to engage China on a regular basis.

To foster greater Singapore-China cooperation in the post-Lee Kuan Yew era, it may be even more important for a younger generation of Singapore leaders to get to know their Chinese counterparts better through such a mechanism, while continuing to draw lessons of experience and insights while collaborating on these projects. At the same time, such a mechanism enables both sides to proactively explore new areas of cooperation to ensure that their relationship remains relevant to their development needs.

Chapter 2

Adaptation, Growth and Innovation

Introduction

On many occasions, the Chinese side has highlighted the stellar socioeconomic progress the SIP has made since 2001 when they started to assume responsibility for the project. In doing so, they tend to overlook the difficulties faced by the Singapore side before they took over. In a sense, the progress made by SIP indicates how the Chinese side has largely fulfilled the challenge as posed to them by the late Lee Kuan Yew. Lee had said then that investors were the best judge of whether SIP was doing well.

Today, SIP stands out from other industrial parks in China for its not only economic achievements but also liveable environment and progress in innovation. The successful industrial take-off in SIP has largely helped to fuel demands for more social services and more resources for social development. To stay ahead of the intense economic competition, SIP would need to constantly restructure and upgrade, which is not an easy task as it is progressing from a high stand-point.

In line with the natural laws of economic progression, SIP is facing the "hollowing out" phenomenon as unprofitable lower-end industries are forced to move out due largely to rising costs and increasing wages. In their place are the more knowledge-based and capital-intensive industries. SIP's ability to attract more high-tech and value-added industries can be attributed to the pro-active role of the local authorities. They have identified biomedicine, nanotechnology and cloud computing as the three new sources of growth. To promote these sectors, the local administrative committee has focused on creating the right software environment for businesses to thrive, notably, through promoting innovation, establishing ecosystems and attracting high-value human resource talents. As in any industrial policy, while some preliminary progress has been made, it is still early to judge whether these three sectors will be "winners" in the long run.

On the social development front, there are areas where the local authorities have not only adapted the Singapore experience but made significant innovations. For one, the neighbourhood centres in SIP appear to have gone a step further by providing more integrated and varied services and amenities under one roof. SIP has further succeeded in bringing rural communities in urban areas under a unified urban administrative system, thus avoiding the pitfalls of an urban-rural divide prevalent in many cities across China. In its drive to create "smart cities" in SIP, the local authorities have further made available to residents a host of social services online. Hence, residents rarely need to appear in person at the service counters of community centres. In this sense, they seem to be more tech-savvy than their Singapore counterparts.

There remains room for SIP and Singapore to learn from each other. Many of the socioeconomic challenges that Singapore is facing today could be a useful reference for SIP. Yet, SIP has also made interesting adaptations and innovations that could also be instructive for Singapore.

Value in the Singapore Connection

The Suzhou Industrial Park Administrative Committee (SIPAC) has described the period from mid-1994 to early 2001 when the Singapore consortium was the majority owner and manager of the project as the "planning" and "foundation-laying" phase.[1] Such a description tends to

[1] See SIPAC website at <http://www.sipac.gov.cn/english/zhuanti/20140429yq20zn/zxhz/fzls/> (accessed 20 March 2017).

underestimate or even overlook the differences between the Singapore and Chinese sides at that time.

The Chinese side today tends to gloss over the initial difficulties in the start-up years and highlights mainly the SIP's subsequent successful development, notably its "accelerated development" phase (from 2001 to 2004) and its "transformation and upgrading" phase (from 2005),[2] claiming full credit for its further development.

In a sense, this outcome indicates how the Chinese side had largely met the challenge posed to them by Singapore's late SM Lee Kuan Yew years ago. In September 1999, a few months before the formal switch in majority stake and management control, Lee had said to the effect that a hallmark for SIP's success would be when investors themselves affirmed that the Chinese side was running the park better than the Singapore team.[3]

In fact, Chinese officials in Suzhou today rarely attribute the success of SIP to efforts made by Singapore in the initial years. Nevertheless, they do make it a point to highlight how the SIP stands out from other industrial parks in China primarily because of the Singapore connection. In promoting SIP, SIPAC officials have often capitalised on the Singapore brand name by highlighting elements such as the integrated and long-term master planning, one-stop service for investors and liveable environment for residents that can all be attributed to Singapore's early contributions. They have also portrayed how the SIP has enjoyed and continued to retain strong political support from leaders in Singapore and China.

In fact, Singapore's footprints are clearly visible in SIP's official records. Upon entering the SIP Urban Planning Exhibition Centre

---

[2] <http://www.sipac.gov.cn/english/zhuanti/20140429yq20zn/zxhz/fzls/> (accessed 20 March 2017).

[3] Lee, in his address to a group of Singapore and Chinese officials in the Suzhou City Convention Centre auditorium, had reportedly said, "Not that you say so, but your customers, the investors. They are the ones who will decide if foreign investments will continue to come into the SIP". See "SM Suggests How to Make Suzhou Park a Success", *The Straits Times*, 27 September 1999.

located at the SIP Archives Building,[4] a prominent oil-on-canvas portrait of Lee Kuan Yew and Deng Xiaoping is on display. It is in recognition of the contributions made by these two chief architects to Singapore-China relations and in making possible the SIP collaboration. In fact, the portrait of these two leaders is set against a background of iconic images of SIP such as Jinjihu (a well-known scenic lake) and Ligongdi (a famous entertainment belt on the southern edge of Jinjihu) and Motianlun leyuan (a ferris wheel cum amusement park to the east of Jinjihu).

Immediately following the two leaders' portrait are embossed renditions of remarks made by succeeding generations of Singapore and Chinese leaders on the SIP.

On the Singapore side:

- "SIP is a grand, unprecedented cooperation plan between the governments of the two countries" (by SM Lee Kuan Yew when he visited SIP on its 10th anniversary, 9 June 2004)
- "Vision, friendship, devotion and experience created a successful Suzhou Industrial Park" (by SM Goh Chok Tong when he visited SIP on an official visit, 15 April 2006)
- "What SIP (Suzhou Industrial Park) has achieved is far beyond our expectation" (by Prime Minister Lee Hsien Loong when he visited SIP on a working visit, 12 September 2010)

On the Chinese side:

- "Speed up the construction of Suzhou Industrial Park, and accumulate new experience in developing Sino-foreign economic and

[4] This SIP Urban Planning Exhibition Centre was inaugurated in October 2014. Singapore's Consul-General in Shanghai Ong Siew Gay delivered a speech on that occasion. See "Inauguration Ceremony for the New Suzhou Industrial Park Urban Planning Exhibition Center and Launch of the 20th Anniversary of China-Singapore SIP Commemorative Stamped Postcard", Singapore Ministry of Foreign Affairs website, 24 October 2014, <https://www.mfa.gov.sg/content/mfa/overseasmission/shanghai/yangtze_delta_ engagements/2014/201410/ Press_20141024_2.html> (accessed 2 March 2017).

technological cooperation through mutual benefit" (by President Jiang Zemin when he visited SIP, 12 May 1995)

- "We hope that both sides will apply the successful experience of Suzhou Industrial Park to new fields of cooperation to achieve more results in cooperation, and endeavor to raise the friendship and cooperation between China and Singapore to a higher level" (by President Hu Jintao when he met SM Lee Kuan Yew in Beijing, 20 June 2004)
- "Jiangsu has a good number of development platforms for open economy such as China-Singapore Suzhou Industrial Park, bonded zones, export processing zones, national economic & technological development zones, and new & hi-tech zones, which are making active efforts in connecting with China (Shanghai) Pilot Free Trade Zone so as to play an experimental and exemplary role in open innovation and comprehensive reform" (by Xi Jinping when he was on an inspection visit to Jiangsu, 14 December 2014)

SIPAC and the Suzhou municipal government's response to the passing of Lee on 23 March 2015 provided another indication of how much the Chinese side value the Singapore connection. Two days after his passing, on 25 March 2015, the SIP Office for Adapting Singapore's Experience in collaboration with the Suzhou Singapore Club held a photo exhibition titled "Lee Kuan Yew and Suzhou Industrial Park" to showcase Lee's instrumental role in SIP's development.[5]

Recalling Lee's seminal contributions, Barry Yang Zhiping (then deputy secretary of the CPC SIP Working Committee and former chairman of SIPAC) reportedly said, "Mr. Lee gave us persistent instructions and support at every stage of [SIP's] development. In the future, SIP will follow the originally conceived goals and learn and

[5] "Lee Kuan Yew Mourned in SIP through Photo Exhibition", SIPAC website, 26 March 2015, <http://stdc.sipac.gov.cn/en/?do=News_AscxNewsView_MTk0OQ> (accessed 2 March 2017). Local leaders who visited the exhibition included Wang Xiang, standing committee member of the Communist Party of China (CPC) Suzhou Municipal Committee and secretary of the CPC SIP Working Committee, and Yang Zhiping, deputy secretary of the CPC SIP Working Committee and chairman of SIPAC.

adapt the successful experience of Singapore selectively to our practice, and translate the ideas of Mr. Lee Kuan Yew into reality. That is the best way to honor the memory of Mr. Lee".[6]

Yang's remarks indicate that SIPAC values and wants to retain what is arguably SIP's biggest selling point, the Singapore connection and the benefits associated with such a linkage, for as long as possible. Seen in this light, it is understandable why SIPAC has retained the 80-sq km Singapore-China cooperation zone in the SIP (out of a total size of 288 sq km).[7] SIPAC has also continued to send its officials to Singapore for software training, a process that has continued unabated since 1994. To date, about 3,400 Chinese officials have visited Singapore under the SIP software programme for training and two-way sharing of development experiences.[8]

Since 2003, the year the joint venture company or CSSD turned in a profit to recover its cumulative losses from the previous years, there have been numerous talks that the company would go for a public listing. This has yet to materialise. One speculation is that the Chinese side is holding back on a listing due to concerns that the Singapore consortium would sell off their shares, thereby further diluting the Singapore component in the SIP.

Growth, Transformation and Upgrading

According to SIPAC, the SIP has undergone four distinct development phases since 1994. They are the (i) planning phase from early 1992 to

[6] "He Leaves Us a Legend: Lee Kuan Yew and Suzhou Industrial Park", SIPAC website, 24 March 2015, <http://www.sipac.gov.cn/english/Topstory/201503/t20150324_346770.htm> (accessed 2 March 2017).

[7] The size of this Singapore-China cooperation zone was originally 70 sq km. It was announced in August 2006 that this cooperation zone would be expanded by an additional 10 sq km, making it 80 sq km.

[8] "13th Joint Council for Bilateral Cooperation (JCBC) Meetings in Beijing, The People's Republic of China", Prime Minister's Office website, 27 February 2017, <http://www.pmo.gov.sg/newsroom/13th-joint-council-bilateral-cooperation-jcbc-meetings-beijing-peoples-republic-china> (accessed 28 December 2017).

April 1994;[9] (ii) foundation-laying phase from May 1994 to June 1999;[10] (iii) accelerated development phase from March 2001 to June 2004;[11] and (iv) transformation and upgrading phase from year 2005 till the present.

Implied in these different stages of development is that SIP only took off from 2001 when the Chinese side became the majority owner and bore responsibility for the project. This suggests that it was the Chinese side that had made possible the accelerated development phase and, more importantly, the subsequent transformation and upgrading phase which is currently on-going.

Accordingly, SIPAC today is very proud of the economic achievements of SIP. These include its growing annual contribution to Suzhou's GDP (gross domestic product),[12] which amounted to 215 billion yuan in 2016 or a 12 times increase from the 18 billion yuan in 2001. Similarly, its annual contribution to Suzhou fiscal revenue at 28.8 billion yuan in 2016 marked a 24 times increase from the 1.2 billion yuan in 2001.[13]

---

[9] Early 1992 refers to Deng Xiaoping's *Nanxun* or Southern Tour where he singled out Singapore as a reference model for China's development. April 1994 refers to the decision by the people's government of Jiangsu province to approve the rural district of Loufeng and the four townships of Kuatang, Xietang, Weiting and Shengpu to be under the direct administration of the Suzhou municipal government (where SIP is located). See SIPAC website, <http://www.sipac.gov.cn/english/zhuanti/20140429yq20zn/zxhz/fzls/> (accessed 20 March 2017).

[10] In May 1994, a "China-Singapore Equity Joint Venture Contract" was signed that called for the creation of a preparatory working group to *inter alia* complete the preparation of a joint venture company to develop the SIP. In June 1999, an MOU (memorandum of understanding) was signed that called for the switch in equity share between the Singapore and Chinese sides by 1 January 2001. By January 2001, the Chinese consortium would own 65% while the Singapore consortium the remaining 35%.

[11] In March 2001, the Suzhou municipal party committee and government held a meeting to mobilise efforts to accelerate SIP's development and construction, signalling the start of the accelerated development phase. In June 2004, SIP celebrated its 10th anniversary.

[12] In China, the practice is to use the term GDP to describe the total value of output produced by a locality even though this term is more appropriately used to describe the total value of output produced at the national level.

[13] "Brief Introduction to SIP", SIPAC website, <http://www.sipac.gov.cn/zjyq/yqgk/201703/t20170317_541391.htm> (accessed 20 March 2017).

Another oft-repeated official line is that though SIP only occupies 3.4% of land and comprises only 7.4% of the population in Suzhou, it contributes about 15% share of Suzhou's GDP.

As a model of sustainable development, SIPAC has stated that the standard carbon emission in SIP per 10,000 yuan of GDP is one-third the national average. The rate of chemical oxygen demand (COD) and sulphur dioxide (SO_2) emission in SIP is only one-eighteenth and one-fortieth the national average respectively.[14] Suzhou, with SIP being an integral part of it, was conferred the Lee Kuan Yew World City Prize in 2014.[15]

The perennial challenge facing SIPAC today is how to ensure that the SIP is constantly moving up the value-chain through restructuring and upgrading so as to stay ahead of the intense competition.[16] In fact, a constant refrain in the research team's conversations with SIPAC is that SIP cannot be contented with its current achievements and rest on its laurels.

In particular, Yang Zhiping (former deputy secretary of SIP Working Committee and former chairman of SIPAC) revealed that SIP had been ahead of the national curve when it embarked on industrial upgrading and innovation more than 10 years ago, way before the current emphasis by the central government. He likened this pioneering and on-going task to the continuous efforts by the Singapore government to identify new areas of growth in light of regional and global developments.[17]

Yang elaborated that driving the SIP towards a high-tech and high-value added economy was no easy task given its already high stand-point.

---

[14] SIPAC website at <http://www.sipac.gov.cn/english/zhuanti/jg60n/gjlnbtsj/> (accessed 20 March 2017).

[15] "Lee Kuan Yew World City Prize", Urban Redevelopment Authority, 2014, <https://www.leekuanyewworldcityprize.com.sg/features_city_focus-Suzhou.htm> (accessed 20 March 2017).

[16] "Transformation a Continuous Process at Suzhou Industrial Park", SIPAC website, <http://www.sipac.gov.cn/english/zhuanti/20140429yq20zn/mtjj/201409/t20140922_295372.htm> (accessed 20 March 2017).

[17] Remarks by Yang Zhiping over lunch that he hosted for the EAI research team on 7 November 2016.

SIP had to grapple with the "hollowing out" phenomenon as unprofitable lower-end industries were forced to move out due largely to rising costs and wages. However, these associated losses in GDP were more than made up for by the more knowledge-based and capital-intensive industries setting up shop in SIP.

According to SIPAC, high-tech industries made up more than 66% of total industrial value-added in 2016 (about the same in 2015). Emerging industries such as biomedicine, nanotechnology and cloud computing generated a value of 47 billion yuan, 38 billion yuan and 35 billion yuan respectively in 2016, growing strongly at double-digit rates of 24%, 36% and 25% respectively compared to that in 2015. Furthermore, the share of the service sector in Suzhou's GDP was 43.8% in 2016, up considerably from 26% a decade ago in 2006.[18]

To support higher value-added growth, SIPAC is cognisant of the importance of having the requisite human resource capabilities. It implemented various talent schemes to attract high-level manpower, of which the most important is the "Jinji Lake Double Hundred Talents Scheme" introduced in 2010.[19] SIPAC has also introduced various social policies and incentives related to housing benefits, living allowances, income and health-care benefits in order to attract the required talent pool from Jiangsu and nearby provinces as well as from overseas Chinese returnees.

Historically, the Greater Suzhou region topped China in producing scholars and this was in fact one of the key considerations for Singapore to site the industrial park in Suzhou. Leveraging on this historical advantage and building on the early master plan, SIPAC has created the Dushu Lake Science and Education Innovation District (*du shu hu ke*

---

[18] "Brief Introduction to SIP", SIPAC website.

[19] In 2010, the "Jinji Lake Double Hundred Talents Scheme" was introduced with the initial goal of attracting/nurturing 200 innovative and entrepreneurial leading talents and 200 high-skilled leading talents every year between 2010 and 2015. This scheme was later renewed and upgraded in 2015. On the basis of attracting/nurturing 200 innovative and entrepreneurial leading talents and 200 high-skilled leading talents in 2015, the upgraded scheme aims to increase the number by 20% every year in the next three years (see Chapter 6 of this book).

jiao chuang xin qu) which today houses not only 28 institutions of higher learning but also various innovative platforms such as the Suzhou Institute of Nano-tech and Nano-bionics (of the Chinese Academy of Sciences), Nanopolis, Biobay, SIP Creative Industry Park and Ascendas iHub.[20]

SIPAC encourages tie-ups between industries and institutions of higher learning in SIP to *inter alia* better tailor manpower development to the needs of industries. This will also facilitate R&D activities and bring products from their experimental stage to the prototype and even commercialisation stages. In the interviews by the research team, companies like Hitachi High-Tech (Suzhou) and Lilly (Suzhou) stressed the importance of nurturing and attracting talents from such institutions for the long-term viability of their operations in SIP.

In a way, the recent spurt in the growth of high-tech and knowledge-intensive industry (which critically depends on the availability of high-skilled manpower) in the SIP has finally borne out Lee Kuan Yew's long-term expectation and early wisdom. He believed that Suzhou in general and SIP in particular would eventually make it in this way because of its strong comparative advantage in education and training.

At the same time, SIP's continued economic dynamism cannot be separated from its social vibrancy. The more liveable SIP becomes, the more attractive it will be in attracting and retaining talent and for their families to settle down in SIP. Being economically dynamic and socially vibrant is hence mutually reinforcing. So is the continued success of SIP to better Singapore-China relations.

The general public's perception of SIP in Singapore remains very much at its early periods when the Singapore side faced difficulties in developing the SIP. In contrast, the Chinese side has largely looked beyond this "unhappy" episode and is more preoccupied with building on the achievements that SIP has made thus far.

---

[20] The Dushu Lake Science and Education Innovation District started in 2002 and has a designated area of 25 sq km.

Social Development and Social Innovation

From its very beginning, the master plan drawn up for SIP had incorporated the development of industrial, residential and commercial sectors, and even green spaces, offering a long-term, integrated approach to developing the SIP.

The priority of the Singapore consortium, when it was the majority owner and manager, was to get the industrial area up and running by attracting international investors and creating jobs. With employment, more residents would come to the SIP, spurring the development of the residential and commercial sectors. However, things did not go according to this plan in the earlier years.

As mentioned in Chapter I, the local authorities were not fully behind the project and even undercut the joint venture company's efforts to attract investors to SIP. An equally, if not, more important reason was that the local authorities dragged its feet on decanting population from Suzhou old city to SIP to promote its growth and development. This foot-dragging stalled the development of the residential and commercial sectors.

In 2001, when the local authorities assumed overall responsibility for SIP's development, it made a more concerted effort to ensure the industrial park's all-round development including its social development. To a large extent, the successful industrial take-off in SIP has fuelled demands for more social services and generated more resources for social development. Today, SIP is known for its social progress and liveability apart from its hard economic achievements.

In striking a balance between development and liveability, the SIP stands in stark contrast to other industrial parks in China that are largely driven by the single-minded pursuit of GDP with scant regard for residents' needs or the environment. This outcome, largely embodies the early vision that Lee Kuan Yew and Singapore had of the project at its conception stage.

Today, SIP is one of China's highly liveable urban environments as demonstrated by its clean and safe surroundings, open and green spaces, variety of quality educational institutions, accessible social services, good infrastructure (such as roads, transit lines and public transport) and other modern facilities and amenities.

A deputy director of the Adapting Singapore Experience Office was extremely proud of the fact that SIP was constantly offering fresh experiences such as the Xinguang Tiandi with its new shopping and commercial space for its residents. He also mentioned that Eslite Bookstore or Chengpin Shudian, a renowned Taiwanese lifestyle chain, has chosen SIP for its flagship store in mainland China.[21]

One defining social feature of the SIP which the local authorities have adapted from Singapore and apparently taken it further is in the development of neighbourhood centres. While the concept of providing residents convenient access to their daily needs remains unchanged, the modus operandi and nature of these services rendered demonstrate a high degree of social innovation.

In Singapore, there is generally a clear separation between public facilities such as community clubs/centres, HDB branch offices and HDB retail spaces (under various government ministries/statutory boards) and facilities such as commercial malls built by private entities on the other. Furthermore, community clubs/centres in Singapore are usually located in a separate building and are sited some distance (albeit within walking distance) from the neighbourhood centres.

In SIP, the neighbourhood centre is an integrated concept that encompasses a wide range of amenities and services including community clubs, libraries,[22] retail space, F&B outlets, supermarkets, fresh produce markets, outpatient clinics, kid's club, leisure and entertainment facilities, and banking facilities.[23]

---

[21] Discussion with a deputy director in SIP in April 2016. Xinguang Tiandi that officially opened in June 2015 is a chain of the famous Taiwanese company Xinguang Sanyue Baihuo. Eslite Bookstore saw its grand opening in SIP in November 2015.

[22] An individual can borrow books from one library (in one neighbourhood centre) and drop it off at a different library in another neighbourhood centre. This practice is similar to that of the National Library Board in Singapore.

[23] Only recently, in Singapore, newer neighbourhood centres like the Tampines Hub provide Singapore's first-ever integrated community and lifestyle hub where residents can

To implement this integrated concept, SIPAC established a state-owned company, the SIP Neighbourhood Centre Development Co. Ltd (SIPNC) in November 1997 to oversee the design, planning, construction and management of the neighbourhood centres in SIP. By the end of 2015, the company had established 17 such centres in SIP.[24] It has further replicated over 30 similar centres in other cities and provinces in China.

SIPAC has further made a concerted push to develop "smart communities" (*zhi hui she qu*) in SIP. The research team noticed that the service counters of the community centres received virtually zero or few walk-in customers. Most residents prefer to access a slew of community services online.[25] In this instance, it would appear that Suzhou residents are generally more tech-savvy than their Singaporean counterparts.

Another notable social innovation in SIP is the integration of rural villages with urban communities, thereby bringing them under a unified system of urban administrative management. This stands in contrast to other Chinese cities where "urban villages" (*cheng zhong cun*) are common and administered separately from the urban areas. Former villagers in SIP now receive urban pension and health-care coverage. They have become stakeholders of an urban lifestyle and this has helped to promote social stability and harmony.

---

access a variety of services such as sports facilities, a regional library, community club programmes and amenities, arts programmes and facilities, a hawker centre and retail shops. See "Our Tampines Hub" <https://ourtampineshub.sg/Our-Tampines-Hub/About-Our-Tampines-Hub> (accessed 19 March 2017).

[24] "Dushu Lake Neighborhood Center Opens", SIPAC website, 23 December 2015, <http://www.sipac.gov.cn/english/categoryreport/InfrastructureAndEcology/201512/t20151223_404389.htm> (accessed 19 March 2017).

[25] Residents can access 63 types of community services (such as health care and family planning, social security, employment, old-age care, public assistance, ethnic and religious affairs, and volunteering) plus 20 types of information services (such as map, transportation, tourist attraction, home service, house renting, job information and so on) online (see Chapter 5 of this book). Residents only needed to appear in person in instances where their online application is incomplete or for collection purposes after their online application has been processed.

SIP "Going Out"

As part of its natural progression, SIP has moved to establish mini-SIPs in other parts of Jiangsu and other provinces in China. This "going out" strategy is closely related to SIP's own transformation and upgrading as lower-end or less competitive industries are phased out to other localities. This is a sound strategy from the economic growth and development perspective.

This "going out" strategy also makes perfect political sense. Located in a more developed part of Jiangsu province, the SIP is in an advantageous position to share its development experience with other poorer or inland areas in China. This is in line with Beijing and even provincial efforts to bring more wealth and development to the less well-off areas.

The Singapore brand name is important for the "going out" strategy of the SIP. The industrial park is widely known in China for being well run and its connections with Singapore. Over the years, many Chinese cities and development zones have organised visits to SIP to learn from its successful experience. By going out, the Suzhou authorities often argue that they are replicating SIP's success in other parts of China.

Both the Suzhou authorities and the joint venture company have been actively involved in setting up mini-SIPs in other parts of China. They have invariably stressed that these projects are being carried out on the basis of the experience gleaned from developing the SIP such as its management and governance structure, planning and design, and pro-business, pro-resident and pro-environment emphasis.

In 2006, the Suzhou authorities embarked on their first external venture, namely, the Suzhou-Suqian Industrial Park, in northern Jiangsu, a poorer region compared to southern Jiangsu where the SIP is located.[26] The Suzhou authorities are also playing a lead role in the development of the Horgos Economic Development Zone, an

[26] The Suzhou-Suqian Industrial Park is 13.6 sq km. A Chinese development limited company is responsible for the commercial development of Suzhou-Suqian Industrial Park. See "Brief Introduction to SSIP", SSIP Administrative Committee website, <http://www.ssipac.gov.cn/ParkProfile/ Index.aspx?CategoryID=6E577D5E-C409-44DB-B187-894AA64A15AC> (accessed 6 March 2017).

important gateway city in distant Xinjiang that borders Kazakhstan (since 2011)[27] and the SIP-Xiangcheng District Cooperative Economic Development Zone (since 2012) in Xiangcheng district to the north of the SIP in Suzhou.[28]

The joint venture company CSSD is also playing a lead role in developing the Sutong Science and Technology Park in Nantong in northern Jiangsu (since 2009)[29] and the Suzhou-Chuzhou Modern Industrial Park in Anhui, a poorer inland province to the west of Jiangsu (since 2012).[30] These two projects, ranging from 30 to 50 sq km, are

---

[27] The Horgos Economic Development Zone spans roughly 73 sq km. In May 2011, state-owned enterprises in SIP formed the Horgos Suxin Real Estate Company to build three major projects in the Horgos Economic Development Zone: a Suxin Centre (comprising office space, commercial units and service apartments), a Suxin Industrial Compound (providing standard industrial facilities mainly for small and medium-sized enterprises) and a Suxin Commune (comprising workers' dormitories). The Suzhou authorities have also organised promotional trips for investors to set up shop in the Horgos Economic Development Zone.

[28] The SIP-Xiangcheng District Cooperative Economic Development Zone spans an area of 47.8 sq km with industrial, commercial and residential components. The Suzhou Industrial Park Urban Renovation and Development Co Ltd (SIPURD) has invested in the 3E Industrial Park in the zone.

[29] The Sutong Science and Technology Park with an area of 50 sq km is located within the Nantong Economic and Technological Development Area. At the second Singapore-Jiangsu Cooperation Council meeting in Singapore in November 2008, the joint venture company, SIPAC and Nantong Economic and Technological Development Area Administrative Committee signed a Letter of Intent to jointly develop an ecologically protected, integrated industrial zone and modern new town, which will combine manufacturing, leisure, commerce and living. The master developer of the park is the China-Singapore Sutong Science and Technology Park Development Co Ltd formed in Nantong on 26 May 2009. (The three shareholders of this company are the joint venture company, Nantong Economic and Technological Development Zone Corporation and Jiangsu Nongken Group Co Ltd) On that same day, Singapore's Deputy Prime Minister Wong Kan Seng, Minister of State for Trade and Industry Lee Yi Shyan, Jiangsu's Party Secretary Liang Baohua and Jiangsu's Governor Luo Zhijun were present at a ceremony to mark the official start of the project.

[30] The Suzhou-Chuzhou Modern Industrial Park spans an area of 36 sq km. The master developer of the park is the China-Singapore Suzhou-Chuzhou Development Co Ltd formed in April 2012. The two shareholders of the company are the joint venture

relatively large in scale and involve urbanisation and industrialisation from the ground up.

The joint venture company is also spearheading smaller scale projects of a few square kilometres relating to the upgrading and renewal of existing urban areas in Jiangsu. They include the China-Singapore Haiyu Garden City in Changshu, a county-level city under Suzhou's administration (since September 2013),[31] the China-Singapore Leyu project (since October 2013)[32] and the China-Singapore Fenghuang project (since October 2014),[33] both located in Zhangjiagang, another county-level city under Suzhou's administration.

Given the variety and more importantly, the varying local conditions, it would be reasonable to expect different degrees of progress on these mini-SIPs. A joint venture company official privately confided to the research team that he finds it easier to promote the Suzhou-Chuzhou Modern Industrial Park in Anhui to companies because it has more attractive features like good infrastructure networks than the Leyu project that the joint venture company is also responsible for.

---

company (56%) and Chuzhou Urban Construction Investment Co Ltd (44%). The park has industrial, commercial, financial and residential components.

[31] The China-Singapore Haiyu Garden City project involves an area of 1.66 sq km out of a total area of 110 sq km in Haiyu township. The agreement to collaborate on this project was signed in September 2013 by three parties, notably, the joint venture company, Changshu city people's government and Haiyu township people's government. The plan is to create a liveable and environmentally friendly area with supporting industries, facilities and amenities. It is meant to be a demonstration project of how the rest of Haiyu township could be modelled on.

[32] The agreement to collaborate on the China-Singapore Leyu project was signed by three parties — the joint venture company, Zhangjiagang city people's government and Leyu township people's government in October 2013. Like the Haiyu Garden City project, the plan is to create a liveable and environmentally friendly area with supporting industries, facilities and amenities.

[33] The China-Singapore Fenghuang project spans an area of 1.58 sq km. The agreement to collaborate on this project was signed by the joint venture company and Fenghuang township in October 2014.

Some Observations

The SIP has had a head start in its development compared to many other industrial parks in China. In this process, it has internalised many attributes of Singapore's software and practice, making the SIP stand out from other industrial parks in China.

Yet, the local authorities rarely give due public recognition to this Singapore contribution even though they are cognisant of the value of the Singapore connection. To a large extent, this is to be expected as since 2001, they have been primarily responsible for driving SIP's development on both the economic and social fronts.

On the economic front, the foremost challenge of the SIP is to stay ahead of the competition. This will not be easy as the SIP is trying to make further progress from an already high stand-point. In a way, SIP's development closely mirrors Singapore's development as Singapore itself is unlikely to experience a spurt in growth or high growth given the mature state of its economy.

The local authorities in Suzhou, like the Singapore government, are determined to play a pro-active role in driving economic development. Already, they have identified three growth areas of biomedicine, nano-technology and cloud computing to nurture and grow.

Growing these new areas calls for a shift in emphasis. In the early years of SIP's development, the emphasis was on creating good infrastructure such as roads, electricity and water to attract foreign investment. Today, the emphasis is on providing the right software environment for businesses to thrive including stressing innovation, promoting ecosystems and attracting high-value human resource talents.

On the social development front, the SIP has shown that it has not only adapted the Singapore experience but also made notable innovations. These include its neighbourhood centres that provide more variety of services and amenities all under one roof and its novel way of replicating neighbourhood centres by forming a company to drive this process.

SIP has further succeeded in integrating rural communities in urban areas under a unified urban administrative system, thus avoiding the urban-rural dichotomy that had plagued many cities across China. The residents in SIP also appear to be more tech-savvy in terms of their willingness to use online platforms to access a host of community services.

In some of these areas, SIP could offer interesting lessons for Singapore. In other areas, many of the socioeconomic challenges that Singapore has faced could still be a good reference for SIP. Thus, there seems to be room for SIP and Singapore to learn from each other as they make progress.

Chapter 3

Tracing the Journey of Growth and Transformation*

Introduction

SIP, the first Singapore-China G-to-G project, is widely recognised as a leading industrial park in China. From a slow start in 1994 with a GDP of only one billion yuan, its GDP had jumped manifold to 215 billion yuan in 2016.

Since 1994, SIP's development can be broadly divided into four phases, namely, the start-up phase of 1994–2000, the economic take-off phase of 2001–2008, the global financial crisis adjustment phase of 2009–2011 and the "New Normal" phase from 2012. The original business model of SIP was to leverage on its low-cost labour coupled with Singapore's experience in economic management and public administration to attract multinational companies to establish export-oriented operations in SIP. This business model worked well in the first two phases of SIP's economic development. The sharp rise in wages and business costs in SIP in the 2000s has since rendered such a model uncompetitive.

Over the years, SIPAC has played a leading role in SIP's industrial upgrading and transformation. This role has become increasingly more

* This chapter is based on a report that has been released as an *EAI Background Brief*, "Suzhou Industrial Park: Innovating With Biopharmaceuticals, Nanotechnology And Cloud Computing Industries (IV)", an EAI team effort. The first draft was prepared by Dr Henry Chan and Ms Ping Xiaojuan.

challenging given the competition posed by other industrial parks in China and SIP's high starting point. Some key transformation trends such as moving from low-end, labour-intensive manufacturing to higher value-added activities are quite apparent, mirroring developments at the national level.

SIPAC is currently focusing on growing the three high-tech industries of biopharmaceuticals, nanotechnology and cloud computing. Unlike the earlier export-oriented industries, these three high-tech industries are largely geared towards the domestic market and most of the companies are of domestic origin. The three high-tech industries are thus less susceptible to the "foot-loose" behaviour of multinational companies of the early periods. The drivers of these industries are mainly overseas returnee and home-grown scientists.

Apart from creating a conducive business environment for further transformation, SIPAC has also made efforts to create a liveable environment in SIP in order to attract high-skilled talents to drive its next phase of growth. SIP will continue to face factory closures and relocation, a natural order of economic progression as the less competitive ones are being phased out to make room for new growth sectors. However, according to SIPAC, the value of companies exiting SIP has been more than made up for by the entry of higher value-added economic activities. It would thus appear that SIP is faring well so far amid a highly competitive environment.

SIP since Its Inception

SIP is in Jiangsu, an important coastal province of China. One of Singapore's early considerations for siting the industrial park in Suzhou was to tap on the city's geographical location as one of the key pivots to China's booming Yangtze River Delta region and its historical advantage in attracting and nurturing talents, which continue to stand the industrial park in good stead today.

In fact, the SIP was one of the three major development zones set up in Suzhou in the 1990s. The SIP has generally performed better than the other two zones, namely, the Suzhou New District High-Tech Development Zone (SND) and Kunshan Development Zone (KDZ).

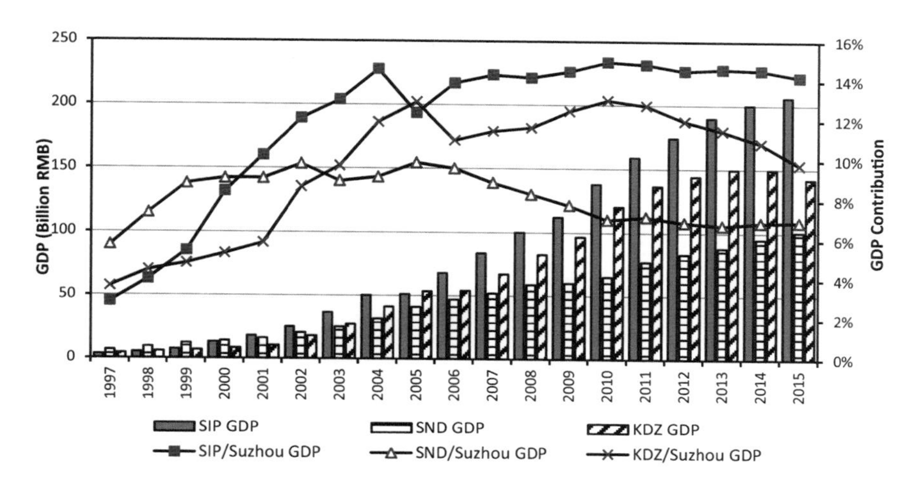

Figure 1. GDP of Three Major Development Zones in Suzhou (1997–2015)
Source: *Suzhou Statistics Yearbooks*, 1998–2016.

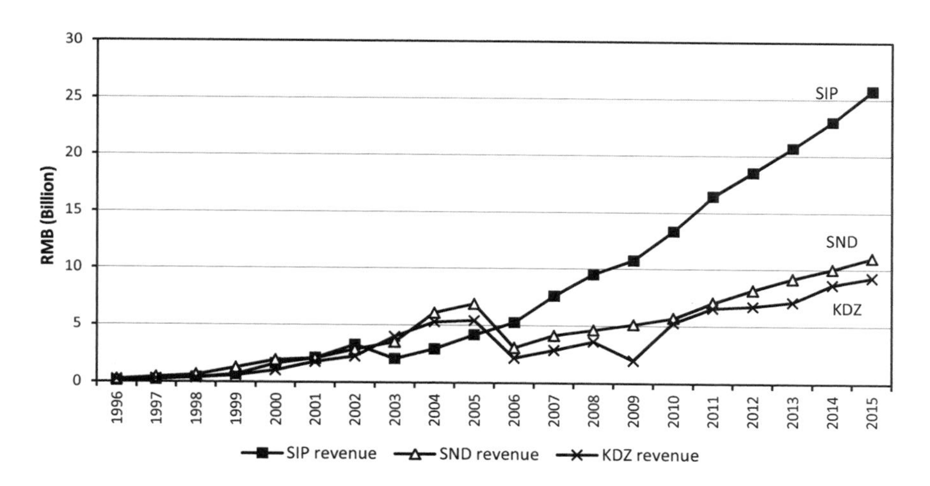

Figure 2. Revenue of Three Major Development Zones in Suzhou (1996–2015)
Source: *Suzhou Statistics Yearbooks*, 1997–2016.

In 2015, SIP contributed almost 15% to Suzhou's GDP, much higher than the contribution of 9.8% by KDZ and 6.9% by SND (Figure 1).

SIP contributed even more to the local revenue base than SND and KDZ. In 2015, SIP generated 25.7 billion yuan, more than double the figure of 11 billion yuan by SND and nine billion yuan by KDZ (Figure 2).

Launched in 1994, SIP operated on a business model that sought to combine Suzhou's low-cost but relatively high-quality labour with Singapore's experiences in economic development, especially in attracting foreign direct investment (FDI) and public administration. In the late 1990s and early 2000s, many multinationals were attracted to set up export-oriented businesses in SIP under the Singapore brand name. The spectacular growth of exports in SIP during the 2000s attested to the success of this model.

However, this original model became increasingly unworkable by the late 2000s. At the national level, rising wage and increasing business costs accelerated with China's economic and export boom following its accession to the World Trade Organisation (WTO) in 2001. The SIP was not immune to this national trend and was under pressure to transform.

Based on the study team's observation, SIP's economic development over the years can be broadly divided into four distinct phases: Phase I — start-up between 1994 and 2000 with its central focus on marketing SIP and infrastructure development and related activities; Phase II — industrial take-off from 2001 to 2008; Phase III — relative deceleration in GDP growth between 2009 and 2011 due to the global financial crisis (GFC); and Phase IV — the "New Normal" of China's economic growth with SIP's GDP growth further coming down to a mid-single digit since 2012 (Figure 3 and Table 1).[1]

Compared to Jiangsu's provincial and China's national GDP growth rates, SIP had performed considerably better during the first three phases, albeit with declining rates of growth. The surge in wages and business costs increasingly rendered its original export-oriented, low-cost labour business model less viable. The sharp deceleration in SIP's average annual growth rate under the "New Normal" highlights the pressure to transform from the earlier model (Table 1).

The same four phases are discernible in SIP's export performance: (i) a slow start in export growth during the start-up phase from 1994

[1] "City Life at Your Fingertips — Report on Fifteen Years of New Village Development in Suzhou Industrial Park" (cheng shi sheng huo chu shou ke ji — 15 nian yuan qu xin nong cun fa zhan diao cha bao gao), 24 April 2009, <http://www.sipac.gov.cn/tbtj/200904/t20090424_42541.htm> (accessed 30 March 2017).

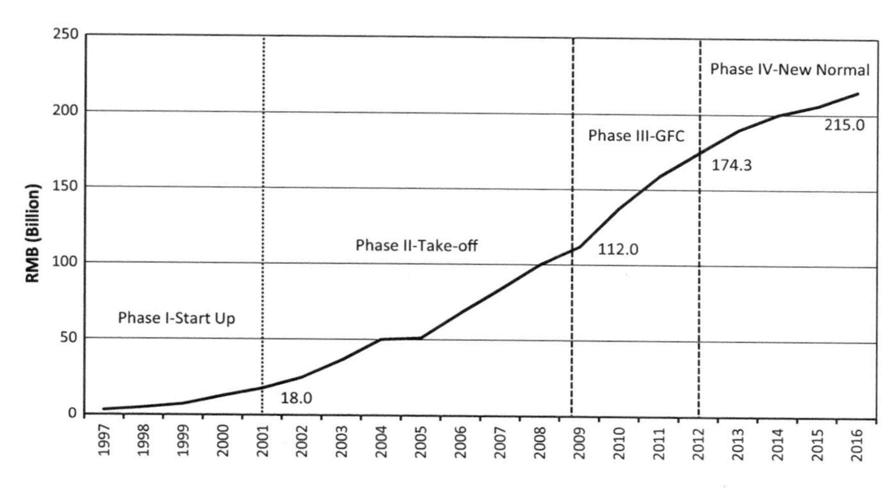

Figure 3. SIP's GDP in Four Phases of Growth (1997–2016)

Source: *Suzhou Statistics Yearbooks,* 1998–2015, meeting notes with Yang Zhiping (then deputy secretary and chairman of SIPAC) and "Introduction of SIP", SIPAC website, <http://www.sipac.gov.cn/zjyq/yqgk/201703/t20170317_541391.htm> (accessed 12 March 2017).

to 2000; (ii) export take-off from 2001 to 2008 with accelerating export growth, averaging 27.3% a year; (iii) export growth slowed down to an average rate of 7.6% from 2009 to 2011 due to the global financial crisis and its subsequent mild recovery; (iv) the "New Normal" phase from 2012 to 2015, with an average export growth rate of 3.7% as exports largely remained stagnant, or even declined (much as for China as a whole), while overall economic growth had become increasingly more domestic-demand driven and service-oriented (Figure 4).

Since 2001, the actualised FDI in SIP generally went on an upward climb but grew much more slowly or even stagnated since 2007, and plummeted noticeably in 2015. This indicates that SIP has been losing its appeal to foreign companies as an export-oriented manufacturing base (Figure 5). In fact, the decline in FDI and exports over the years is closely interrelated.

The primary cause behind the recent decline of both exports and FDI is the sharp rise in wage and business costs in SIP. During an earlier

Table 1. Average Annual Growth Rate of SIP Versus Jiangsu and China

		Average Annual Growth Rate		
	Period	**SIP**	**Jiangsu**	**National**
Phase I: Build-up	1994–2000	More than 40.0% (see Note 1)	13.4%	9.5%
Phase II: Take-off	2001–2008	29.8% (see Note 2)	17.5%	10.7%
Phase III: Global Financial Crisis (GFC) and Recovery	2009–2011	16.7% (see Note 3)	16.6%	9.8%
Phase IV: "New Normal"	2012–2016	6.2% (see Note 4)	9.3% (2012–2015)	7.4%

Source: *Suzhou Statistics Yearbooks*, 1998–2016, *China Statistics Yearbook*, 2016 and meeting notes with Yang Zhiping.

Note 1: Attracting investors and infrastructure build-up from a low base accounted for the high growth in Phase I.
Note 2: Export-oriented manufacturers set up shop and SIP's export-oriented production took off in Phase II.
Note 3: Residual momentum from existing businesses keep growth high but output growth slowed down in Phase III relative to the earlier take-off phase.
Note 4: High wage and business costs forced low valued-added companies in SIP to relocate and SIP upgrades by attracting high-tech industries in Phase IV.

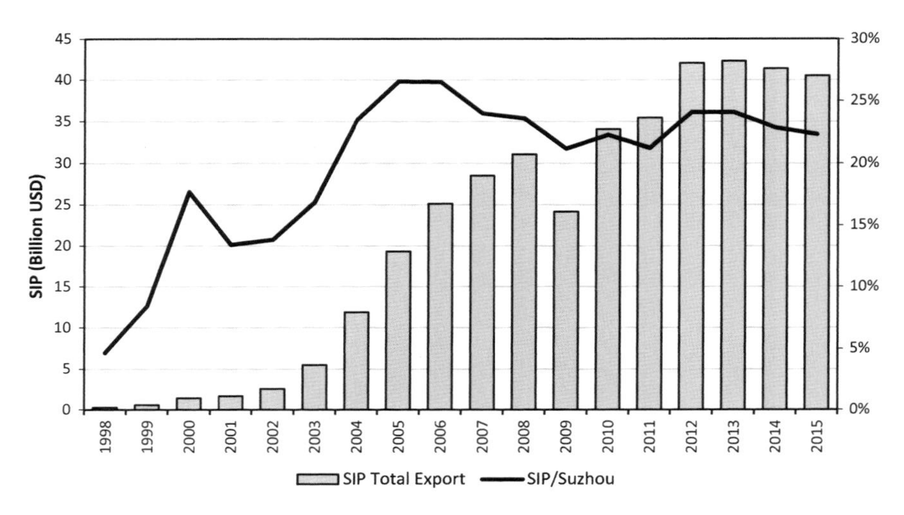

Figure 4. SIP's Share in Suzhou's Total Exports (1998–2015)

Source: *Suzhou Statistics Yearbooks*, 1998–2016.

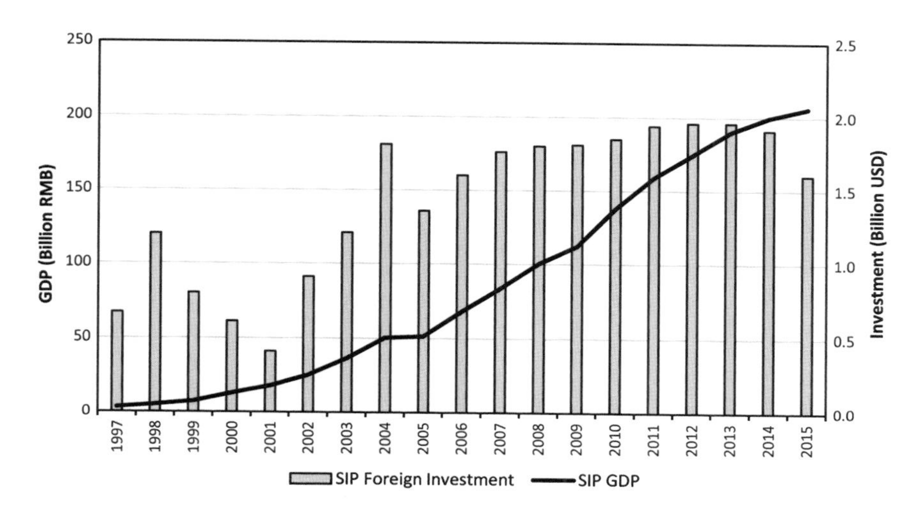

Figure 5. GDP Growth and Actualised FDI in SIP (1997–2015)
Source: *Suzhou Statistics Yearbooks*, 1998–2016.

EAI study trip to SIP in 2009, some companies had already highlighted the high costs of operating in SIP.[2] According to official figures, average annual wages had risen by more than 450% between 1999 and 2016, from 16,456 yuan to 90,573 yuan (Figure 6).

In a sense, SIP's wage increase pattern corresponds with the national trend in general and coincides with policies/initiatives undertaken at the national level. From 2009 to 2011, during the global financial crisis, the rate of average annual wage increase in SIP rose markedly to 14.5%, a reflection of the central government's huge stimulus package of four trillion yuan in November 2008 and rapid credit expansion in 2009–2010 in order to pump-prime the national economy (Table 2).

These measures while meeting the short-term goal of sustaining and even raising production eventually forced wage and business costs to go up for all export-oriented enterprises in SIP. At the same time, the

[2] John Wong, Zheng Yongnian and Lye Liang Fook, "Suzhou Industrial Park: The Singapore Experience (II)", *EAI Background Brief*, No. 451, 13 May 2009, East Asian Institute, National University of Singapore (restricted circulation).

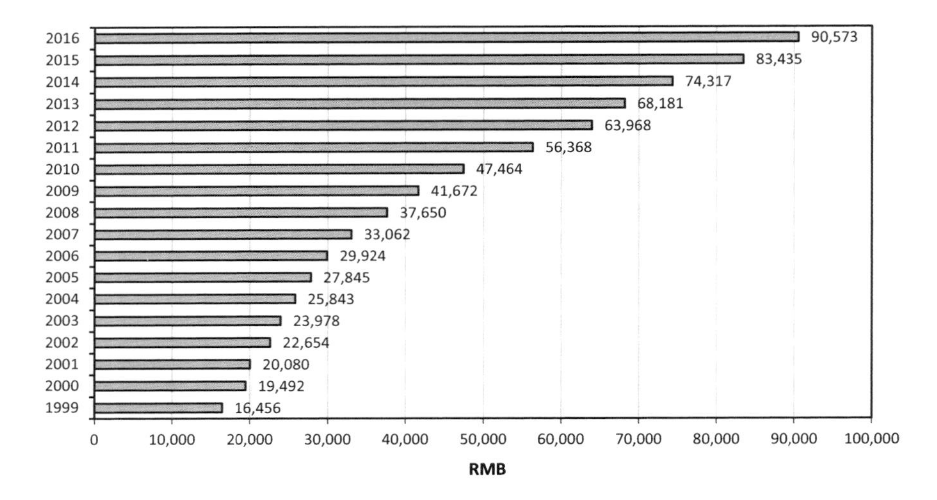

Figure 6. Average Annual Wage Hike in SIP, 1999–2016

Source: *Suzhou Statistics Yearbooks*, 2000–2016.

Table 2. Average Annual Wage Increase of China and SIP

	Average Annual National Wage Increase	
Period	**National (%)**	**SIP (%)**
2001–2008	15.2	8.6
2009–2011	13.1	14.5
2012–2015	10.4	9.7

Source: *Suzhou Statistics Yearbooks*, 1998–2016 and CEIC database.

Note: Computation based on simple arithmetic mean of annual wage increase in the designated period.

global financial crisis that created deflationary pressures worldwide had further exerted a downward pressure on the price of goods of export-oriented enterprises in SIP. These enterprises were therefore squeezed by rising costs and slimmer profit margins.

Industrial Upgrading Pressures

A major preoccupation of SIPAC, the local authority responsible for the overall development of SIP, is to focus on industrial upgrading to ensure that the SIP stays competitive all the time. With stagnating or declining exports and foreign investments in recent years, the pressure to upgrade has become much more urgent.

SIP used to emphasise the promotion of three export-oriented sectors: (i) electronics, software and R&D; (ii) mechanical and precision engineering; and (iii) chemical, health care and pharmaceuticals. These companies were at that time, largely involved in downstream manufacturing or the assembling end of these three sectors, nominally high-tech but relatively low value-added.

According to a senior SIPAC official, SIP has been losing 20 to 25 billion yuan in annual production value in the recent years as some existing companies face cost pressures to relocate to other less developed Chinese cities or even overseas.[3] Rising wage cost is one of the key reasons for the relocation.

One high-profile case was the closure of Seagate Technology (Suzhou) in SIP in January 2017 with almost 2,000 workers reportedly being laid off. In 2015, the company had exported US$1.2 billion and imported US$720 million worth of goods and was considered one of the major exporters in SIP. It has since relocated to Thailand.[4]

The economic rationale behind Seagate's departure from SIP appears to be the same as the closure of its factory in Ang Mo Kio (Singapore) in 2000. SIP is thus experiencing the same industry migration pressure — cost pressures such as wage increases which have rendered certain production activities unprofitable — faced by Singapore much earlier.[5]

---

[3] Interview with the director of Research Office, SIPAC, 19 April 2016.

[4] The Seagate plant was one of the well-known factories in SIP in the 2000s. The factory started operations in SIP in 2003 with a factory floor area of more than 100,000 sq m.

[5] From 1986 to 1996, Singapore accounted for up to half of the hard disk drives produced worldwide and the employment of 80,000 workers at its peak. In the mid-1990s, Seagate

Samsung Electronics, another well-known company in SIP, scaled down its presence in 2016 by closing its large television production plant. Invariably, following such closures, SIPAC would have to find alternative uses for the vacated sites so as to generate revenue for SIP. Samsung Electronics' vacated site has been turned into a school.[6]

Underscoring the high company attrition rate or relocation of companies from SIP to other locations, a senior executive of the semiconductor foundry, AMD Technologies (China) Co Ltd, commented that his company is "among the 10 to 15% of companies that have remained in SIP from the earlier years".[7]

AMD Technologies began operations in SIP in 2005 to provide integrated circuit (IC) assembling, testing, marking and packing services to the mother company AMD, a major global IC foundry. The Suzhou plant is involved in the downstream low-value assembling end of the IC production chain. The company folded into a joint venture with the Shenzhen stock exchange-listed Nantong Fujitsu Microelectronics Co Ltd in 2016.[8]

While rising wages and business costs have driven many companies to relocate, some still find it viable to remain in SIP. According to a senior representative of Gultech (Suzhou) Electronics Ltd, wage costs had risen from 6% of operating cost in 2000 to 20% in 2016. Despite its profit margin being significantly squeezed, the representative said that Gultech, not being labour-intensive, would stay put for the moment mainly due to the value-chain it has established with upstream and downstream businesses in SIP (Box 1).

---

was the second-biggest private sector employer in Singapore with 18,000 workers. All hard disk drives manufacturing activities ceased operation in Singapore by 2014.

[6] Interview with the marketing director, Linray Investment, 8 November 2016.

[7] Interview with the corporate vice president operations, and managing director, AMD Technologies (China) Co Ltd, 20 April 2016.

[8] "AMD and Nantong Fujitsu Microelectronics Co Ltd. Close on Semiconductor Assembly, and Test Joint Venture", AMD website, 29 April 2016, <http://www.amd.com/en-us/press-releases/ Pages/amd-and-nantong-2016apr29.aspx> (accessed 30 June 2017).

Box 1. Gultech (Suzhou) Electronics Co Ltd

The company is a subsidiary of Gul Technologies Singapore Ltd (Gultech for short). Gultech started out as a US-based Data General in-house captive printed circuit board (PCB) plant in Singapore in 1984 covering the Asia-Pacific market. Gultech was founded via management buy-out in 1988; it was listed on SESDAQ in March 1997 and subsequently transferred to the main board of SGX in July 2000. Gultech delisted from SGX in January 2013. The Singapore main board listed Tuan Sing is the current major stockholder with 44.5% of the company's shares.

The company became a wholly owned subsidiary of Gultech on 31 December 2015 when Gultech bought out the 38.6% minority shareholder of the company, Anhui Cord Fabrics Company Ltd, at 194 million yuan. Gultech is a key player in the global PCB industry for the following six markets: automotive, computer and its peripherals, consumer electronics, telecommunication, health care, instrument and control. The mother company has three manufacturing facilities in Jiangsu province, namely, Gultech (Suzhou), Gultech (Wuxi) and Gultech (Jiangsu).

Gultech (Suzhou) started operating in SIP in 2000; its current production capacity is 500,000 sq ft of PCBs a month. It has moved upscale from the original double side to 14-layer board production. The Gultech (Wuxi) plant started operation in October 2004 and can now produce 800,000 sq ft of PCB of up to 20 layers, more advanced HDI (High Density Inter-connected) PCBs and rigid-flex PCBs. The Wuxi plant also houses an R&D division. The Gultech (Jiangsu) plant is located near Gultech (Wuxi) and will eventually produce 1,200,000 sq ft of PCBs after completing four phases of expansion. The Gultech (Jiangsu) plant started operations in April 2015.

Source: GUL Technologies manufacturing facilities, <http://www.gultech.com/ManuFaci.htm> (accessed 30 March 2017).

Another challenge faced by some companies is the stringent environmental standards that SIP has set. A Gultech (Suzhou) representative cited difficulties in expanding its operations in SIP due to the high environmental standards enforced. For instance, SIPAC requires the company to solicit public opinion for any of its expansion plans, essentially putting its expansion plan on hold.

Elaborating further, the representative explained that the public is unlikely to approve of any expansion plans for fear of toxic wastes generated from the production of PCBs in the neighbourhood.[9] Given this constraint or what is commonly known as the NIMBY (Not In My Back Yard) Syndrome, the company appears to be banking more on its Wuxi operations to meet its expansion plans. Thus, it could just be a matter of time before Gultech (Suzhou) would move out of SIP.

Towards More Innovative Type of High-tech Industries

In the mid-2000s, SIPAC started promoting the growth of three high-tech industries: biopharmaceuticals, nanotechnology and cloud computing.[10] As opposed to the earlier focus on relatively low value-added industries, the current three high-tech industries emphasise creating the entire value chains,[11] particularly for their upstream high value-added R&D components.

---

[9] Interview with the vice president, Gultech (Suzhou) Electronics Co Ltd, 9 November 2016.

[10] In March 2017, SIPAC started to use "Artificial Intelligence and related industries" to refer to "Cloud Computing and related industries".

[11] High-tech industry value chain involves upstream R&D, prototype, manufacturing/assembling and marketing. In the case of high-tech consumer products, upstream R&D and end stage marketing often comprise the lion's share of value creation, hence the term "smiling curve". In the case of high-tech industrial goods, marketing normally does not involve much value creation. Rather, it is the upstream R&D component that generates the most value. SIPAC earlier focus in the mid-2000s on promoting certain industrial clusters did not appear to put much emphasis on upstream R&D activities.

These three more innovative high-tech industries have witnessed robust growth. They recorded total sales of 47 billion yuan (biopharmaceuticals), 38 billion yuan (nanotechnology) and 35 billion yuan (cloud computing) in 2016, growing by 57%, 36% and 25% respectively from 2015.[12]

Unlike the earlier export-oriented industries, the current three high-tech industries are largely geared towards the domestic market and most of the companies are domestic ventures. A primary source of human capital involved in such high risk, high value-added, R&D-intensive industries are home-grown and overseas returnee scientists.

In fact, SIPAC recognised early the importance of having the requisite talent pool to drive industrial upgrading in SIP. In 2006, it launched the "10-Year Talent Ecosystem Build-up Programme" (*shi nian gou zhu sheng tai quan*).[13] Earlier in 2002, the Suzhou Dushu Lake Higher Education Town (*du shu hu gao deng jiao yu qu*) was established in SIP. Today, the education town has attracted 26 universities with more than 90,000 enrolments, 72,000 of whom are undergraduates and 18,000 are postgraduates.

Due to its emphasis on higher education, SIP has developed one of the most educated labour forces among the development zones in China. According to official sources, more than 40% of its workforce has completed tertiary education, with 12% holding a master's degree or higher.[14]

---

[12] Interview with SIPAC officials on 7 April 2016. In recent years, these three high-tech industries grew at a rate of 25% to 30% annually.

[13] The "10-Year Talent Ecosystem Build-up Programme" is designed to attract leading entrepreneurs and scientists to relocate to SIP to fill the business entrepreneurship and scientific talent void in SIP.

[14] "Technological Pioneering Talents Project in Suzhou Industrial Park, Constructing 'Ecosystem for Talents' in Ten Years" (yuan qu ke ji ling jun ren cai gong cheng shi nian gou zhu "ren cai sheng tai quan"), 28 October 2016, <http://news.sipac.gov.cn/sipnews/sytt/201610/t20161028_490942.htm> (accessed 30 March 2017).

By May 2017, SIP had attracted 150 overseas returnee scientists under the national "Thousand Talents Programme" (*qian ren ji hua*)[15] and 15 members of the prestigious Chinese Academy of Sciences and Chinese Academy of Engineering are also working in SIP. Such high-end talents are critical to the development of high-tech industries in SIP.

Apart from human resources, SIP has sought to build an ecosystem of manufacturing and venture capital funding for these three high-tech industries. As of 2016, there were 64 incubators, 51 specialised national laboratories and 147 multinational companies R&D facilities in SIP.

More Service-Oriented Activities

Another key aspect of SIP's transformation is the shift from a manufacturing-based, export-led economy to one where services play an increasingly important role. The service industry covers areas that not only focus on innovation and R&D but also sectors like logistics, outsourcing, finance, education, tourism and property development.

Over the years, the share of services had increased from around 20% of GDP in 2000 to about 42% in 2016. Correspondingly, over the same period, the share of manufacturing declined significantly from over 70% in 2000 to around 57% in 2016 (Figure 7). This shift towards more service activities in SIP is in line with the general pattern of structural change — towards more consumption- and service-driven sources of growth — in the national economy.

On the financial services front, Suzhou moved its major financial institutions from the old district to Jinji Lake Central Business District in SIP in the late 2000s. A senior executive of Soochow Securities commented that Jiangsu is the second-largest province in China in terms of regional GDP, registering annual output of more than seven trillion yuan. SIP is in southern Jiangsu, the richest part of Jiangsu, and is a

[15] The "Thousand Talents Programme" is the most prestigious national talent attraction programme administered by the Central Organisation Department of the Chinese Communist Party. It is designed to attract innovative talents under 55 years old in various disciplines who are willing to work in China on a full-time basis.

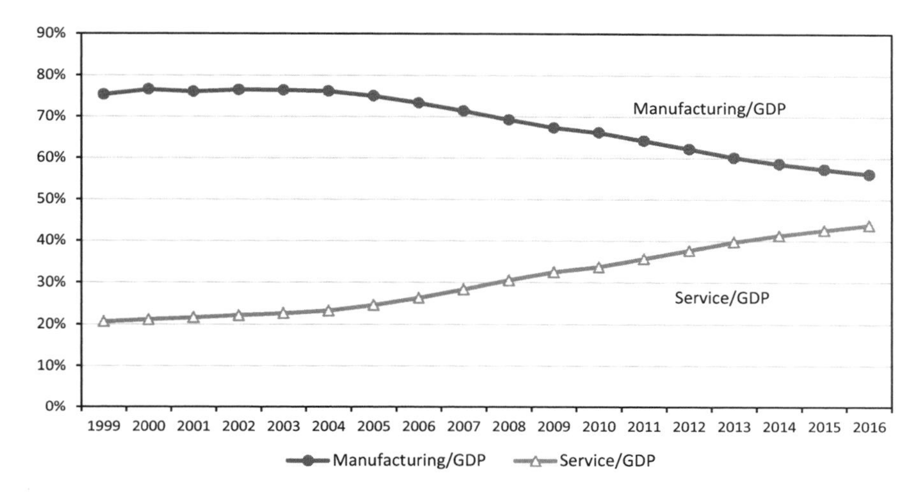

Figure 7. SIP's Increasing Service Share in GDP (1999–2016)

Source: *Suzhou Statistics Yearbooks*, 2000–2016; Data from SIPAC, <http://www.sipac.gov.cn/government/tjfx/ 201703/P020170320341751401534.pdf> (accessed 31 March 2017).

favourite choice of residence for the wealthy from Jiangsu and Shanghai, which makes SIP an ideal place for fund management business (Box 2).[16]

Sounding a cautionary note, the executive from the same company pointed out that SIP is not likely to become a major financial centre due to the prevailing strengths of other localities. In his view, "Beijing has political advantages, Shanghai has a big financial base, while Shenzhen has flexible mechanisms plus proximity to Hong Kong". He noted that financial institutions in SIP had so far not been able to divert financial business away from Shanghai.

As at end 2016, there were 129 licensed financial institutions and a total of almost 800 financial firms including small financial companies, trust and guaranty companies, leasing companies, venture capital (VC) and private equity (PE) investment firms in SIP.[17]

[16] Interview with the vice president of Soochow Securities Co Ltd, 8 November 2016.

[17] Interview with the director of Finance Management and Service Bureau, SIPAC, 7 November 2016

Box 2. Soochow Securities Co Ltd

Soochow Securities is a Suzhou state-owned enterprise (SOE) that was relocated to the SIP in 2009 at the request of the municipal government. The company enjoyed concessional land price and its new headquarters was sited on the west side of Jinji Lake. The company made a second move to its current headquarters on the east side of Jinji Lake in 2013.

Soochow Securities provides a full range of stock broking and asset management services in China. As at end September 2016, it had assets of 80.2 billion yuan, stockholder equity of 20.1 billion yuan and almost 1.4 million customers, and ranked 22nd in China's securities industry. Net income for the first three quarters in 2016 was 3.3 billion yuan.

The company led the SIP's "going out" strategy for local financial institutions in SIP. It has set up an asset management joint venture company with CSSD in Singapore. The Singapore office is the first overseas office for Soochow Securities. This is quite an unusual move as virtually all Chinese financial institutions pick Hong Kong rather than Singapore as their first overseas ventures.

Source: Interview with Li Qibing, vice president of Soochow Securities, 8 November 2016.

Start-ups in SIP provide a fertile ground for VC and PE business. Although SIP may not be able to challenge Shanghai as the financial centre of east China, the expansion of financial services still provides an additional source of growth for SIP and attracts more well-off immigrants.

Another source of growth for the service sector is SIPAC's efforts in promoting urbanisation. SIPAC confers *hukou* or residence permit to well-off migrants for their purchase of residential properties in SIP. In the interviews, the access to good living environment and good educational facilities are often cited as factors for the move to SIP. SIPAC further provides a generous package for talents to settle down in SIP which invariably includes subsidies for their residential property purchases.

SIP controls the number of annual new *hukou* immigrants to avoid overloading the environment and social services in SIP. A criterion for acquiring SIP *hukou* is to buy a residential apartment in SIP. In this way, the *hukou* granting scheme provides another revenue stream for SIP to fund its economic transformation efforts.[18]

The large pool of well-off migrants to SIP provides a stable base for the expansion of the service industry such as in education, health care, banking, retail, and food and beverage. SIP's overall economic growth has thus been steadily restructured towards greater and more sustainable domestic demand-based service activities.

Transformation Challenges

In the interviews, former SIPAC Deputy Secretary and Chairman Yang Zhiping and his officials clearly took great pride in the progress made by the SIP and in its goal to be always at the forefront of economic and social transformation, or in their experimentation with various new economic initiatives, way ahead of other industrial parks at the same or even national level. Yang even stressed the point that well before Beijing raised the importance of innovation and R&D, SIP had already embarked on this area.[19]

For SIP to remain successful and ahead of other industrial parks in China, SIPAC is constantly reviewing SIP's existing industrial structure and identifying new sources of growth. In a way, Singapore's software transfer of the early years is evident in SIPAC's pro-business, pro-active and forward-looking modus operandi.

---

[18] As at end 2013, SIP had 1.028 million permanent residents, of which only 413,000 had *hukou* status. In addition, there were 575,600 migrants. All revenues generated by land sales go to SIPAC. As land price usually accounts for a typical 50–70% of an urban apartment complex in China including in SIP, the industrial park generated significant revenue from property sale.

[19] Interview with Yang Zhiping, SIPAC chairman, 7 November 2016.

SIPAC has successfully managed a number of key economic transitions in SIP. For one, SIP has moved from relatively low-end, labour-intensive manufacturing to high-end and more capital- and R&D-intensive manufacturing. It has also shifted from a heavy reliance on the manufacturing sector to one where services have become an important component of the local economy.

Through the industrial policy of "spotting the winners", SIPAC has identified three high-tech areas, namely, biopharmaceuticals, nanotechnology and cloud computing, as new drivers of growth. They mark another phase of SIP's transition towards high value-added and R&D-intensive industries.

SIPAC has further promoted SIP as a highly liveable place as part of its overall strategy to make the industrial park socially vibrant in order to attract high-skilled and other requisite talents to the park to provide the critical manpower that drives its next phase of growth.

As long as SIP remains socially vibrant, its service sector will continue to grow. A more vibrant community will demand more social services like housing, social security, education, health-care and other modern-day amenities and facilities. SIP's financial sector is also poised to grow, not to rival Shanghai, but to support the growth of the service sector in SIP.

While the SIP is losing 20–25 billion yuan in annual production value due to factory closures and relocations, this loss has been more than made up for by the entry of higher value-added investors in SIP. SIPAC is evidently coping well with the economic restructuring and industrial upgrading so far.

However, future efforts will not be easy as SIP is progressing from a high stand-point of development and further progress is likely to be more difficult and real benefits more incremental. In addition, the industrial policy of "spotting the winners" still poses a certain element of risk as not all that are chosen may become winners in the end.

For instance, a high performing sector with entire value chains, particularly with upstream high value-added R&D components, tends to have wide repercussions once its underlying technology becomes obsolete. A classic example is the decline of Rochester city in the

United States. The city was once home to industrial powerhouse Eastman Kodak but suffered a serious economic meltdown when Kodak's traditional chemical-based photography was decimated by the emerging digital photography.[20]

For SIP to remain competitive and ahead of the intense competition, SIPAC will need to constantly identify opportunities and anticipate challenges. In a sense, what SIPAC is doing is akin to what the Singapore government has all along been doing in terms of keeping the Singapore economy competitive for the future.

---

[20] At its peak, Eastman Kodak reportedly produced 90% of the film used in the United States and provided jobs for more than 60,000 local people. See "Last Days of Kodak Town: the Decline and Fall of the City Photography Built", *The Guardian*, 25 June 2014, <https://www.theguardian.com/artanddesign/2014/jun/25/kodaks-town-decline-and-fall-of-city-photography-built> (accessed 14 July 2017).

Chapter 4

Innovating with New Industries*

Introduction

To transform SIP, SIPAC began to promote three high value-added innovative industries of biopharmaceuticals, nanotechnology and cloud computing in the mid-2000s. As the growth of the three high-tech industries depends heavily on highly skilled talents, SIPAC has not only improved the living environment in SIP, but also provided many incentives such as housing subsidies, tax breaks, R&D grants, venture capital and incubator start-up support. Suzhou's strength in education and skills training is another factor in attracting scientific and technical talents to SIP.

The biopharmaceutical industry focuses on the development of biosimilar drugs which are generic versions of their original counterparts. The focus reduces the product development time of such drugs and minimises the risks involved. The high-skilled talent of this industry is mainly staffed by returning Chinese-overseas scientists. Leading companies in this industry include local champions such as Innovent Biologics and CStone Pharmaceuticals. The industry targets the domestic market and SIP is recognised as a key biopharmaceutical production cluster in China.

---

* This chapter is based on a report that has been released as an *EAI Background Brief*, "Suzhou Industrial Park: Innovating With Biopharmaceuticals, Nanotechnology And Cloud Computing Industries (IV)", an EAI team effort. The first draft was prepared by Dr Henry Chan and Ms Ping Xiaojuan.

The nanotechnology industry in SIP draws on the strength of China in material science research. The Chinese Academy of Sciences set up the Suzhou Institute of Nano-Tech and Nano-Bionics in the early 2010s to conduct both basic and applied R&D at the SIP. Among the three high-tech industries, nanotechnology has secured the most start-up research funding from the Chinese government. The industry is staffed by both returning Chinese-overseas scientists and home-grown Chinese scientists.

The cloud computing industry at SIP used to concentrate on Software as a service (SaaS) dominated by numerous small and medium-sized start-ups. This industry is transforming into a more comprehensive model that includes Platform as a service (PaaS) and Infrastructure as a service (IaaS) with the entry of corporate giants such as Huawei. As the three high-tech industries rely on high-skilled labour and target the domestic market, they tend to be less susceptible to the "foot-loose" behaviour of companies when labour and business costs rise.

At the same time, these industries have created complete value chains: from upstream R&D to downstream manufacturing/assembling end with home-grown talents and returning scientists. The future sales per employee are likely to increase, thereby enabling SIP to further contribute to local GDP. However, as the three high-tech industries emphasise application type of innovation rather than the basic but disruptive type of innovation, such a focus may constrain the growth of these industries over the longer term.

Overall, employment in SIP is not expected to grow much even as production value of the three high-tech industries increases. The continuing relocation of existing low value-added companies will affect overall employment growth.

SIPAC's Critical Role in Industrial Upgrading

As part of its industrial policy of "spotting the winners", SIPAC, which oversees the development of the SIP, has since the mid-2000s been promoting three innovative but domestic market-oriented high-tech industries, namely, biopharmaceuticals, nanotechnology and cloud computing. In fact, SIPAC has provided generous incentives and fostered a favourable ecosystem for these industries.[1]

The favourable ecosystem comprises strong local government support and good physical infrastructure, favourable venture capital environment and readily available scientific/entrepreneurial talents. SIPAC funded such ecosystem build-up by using its extensive state-owned enterprise (SOE) network (Box 1).[2]

SIP further provides generous incentive schemes to develop high-level human resources by attracting both overseas and home-grown scientists. By and large, overseas returnee scientists kick-started the biopharmaceutical industry and a combination of home-grown and returnee scientists worked together to start the nanotechnology industry, while cloud computing depended mostly on home-grown Chinese talents.

Both the home-grown and overseas returnee talents were attracted by SIP's excellent living environment and competitive enumeration.[3] SIPAC has also provided other incentives such as home purchase subsidies, tax breaks, interest-free loans, R&D grants, VC and incubator set-ups for companies in these industries, especially the start-ups (Appendix 1).[4]

---

[1] "Suzhou Industrial Park: Leading the Way to a More Creative Future", GlobeNewswire, 20 December 2016, <https://globenewswire.com/news-release/2016/12/20/899200/0/en/Suzhou-Industrial-Park-Leading-the-Way-to-a-More-Creative-Future.html> (accessed 30 March 2017).

[2] In 2015, the SOEs under SIPAC had a total asset of 189.4 billion yuan, net equity of 62.4 billion yuan, total sales of 19.3 billion yuan and net income of 1.6 billion yuan. Out of SIP's total fixed asset investment of 54 billion yuan, the SOEs accounted for 24% of total fixed asset investment.

[3] Interview with the CEO and board chairman of Kmerit (Suzhou) Information Science and Technology Co Ltd, 9 November 2016.

[4] According to Benjamin Shobert, senior associate, National Bureau of Asian Research of the United States, six basic conditions are required for setting up a vibrant domestic high technology sector. They include government incentives and funding for basic science; talent in the form of both scientists and commercial specialists; a robust linkage between

Box 1. SIPAC's Role behind Venture Capital and Private Equity Funding

The investment holding arm of SIPAC, Oriza Holding (元禾控股), owns a venture capital subsidiary, Heyu Group (禾裕科技金融). As of June 2016, Heyu Group had provided 7,500 venture capital credit support to the tune of 26 billion yuan to more than 400 high-tech start-ups. Oriza Holding is active in the later phase of financing, having invested more than 20 billion yuan as of June 2016.

The active venture capital (VC) and private equity (PE) investment approach of SIPAC is common in China as in other countries, but its scale in terms of amount invested and the number of invested companies is very high as compared to other high-tech industrial parks in China.

Source: "Is 'Going-out' Difficult, Applaudable, or Remarkable" "('zou chu qu shi nan shi hao shi, huan shi niu shi)", *Modern Suzhou* (xian dai su zhou), May 2016, pp. 44–47.

To provide entry-level researchers for R&D activities, SIPAC has made a concerted effort over the years to attract universities and other tertiary educational institutions to set up shop in SIP. This domestic talent source complements and supplements the external talent pool from overseas and elsewhere in China.[5]

---

the government, academia and private sector; a vibrant venture capital space; a market rewarding innovation stakeholders; and a benign environment addressing the long gestation period on innovation and commercialisation.

"Benjamin Shobert Testifies before the U.S.-China Economic and Security Review Commission", The National Bureau of Asian Research, 16 March 2017, <http://www.nbr.org/research/activity. aspx?id=750> (accessed 17 March 2017).

[5] "Suzhou Accounted for 111 Academicians in the Chinese Academy of Sciences and Chinese Academy of Engineering, and Who Are They" (su zhou ji liang yuan yuan shi

By 2015, more than 40% of the workforce (totalling about 300,000) in SIP held college degrees and/or above, with 12% of them holding master's degree and/or above. SIP has thus successfully attracted over 150 overseas returnee scientists/entrepreneurs under the national "Thousand Talents Programme", the most prestigious talent scheme managed by the Organisation Department of Chinese Communist Party Central Committee.[6]

The three high-tech industries have also built up the R&D component to support the growth, with most of the R&D activities done by local Chinese companies or local research institutes. They pragmatically focus on application-type, rather than basic, disruptive innovations.[7] This significantly reduces the potential commercial risk associated with highly original and basic type of R&D.

In short, these three high-tech industries have been largely built up by overseas returnees together with home-grown scientists. These industries are much less dependent on foreign talents from multinationals. In this sense, SIP is less likely to suffer from the "foot-loose" syndrome caused by rising costs, as experienced in many high-tech industrial parks around the world. In terms of talent localisation, these three industries have taken root in SIP.

---

zeng zhi 111 wei du you na xie da wan), Suzhou Mingcheng News, 8 December 2015, <http://news.2500sz.com/news/szxw/ 2015-12/8_2821164.shtml (accessed 23 March 2017). Natives of Suzhou accounted for 111 (or 6.8%). academicians out of a total of 1,629 academicians in the Chinese Academy of Sciences and Chinese Academy of Engineering as at end 2015.

[6] "Suzhou Industrial Park Major Talent Policies in 2017" (2017 su zhou gong ye yuan qu zhu yao ren cai zheng ce hui bian), SIPAC, <http://www.sipac.gov.cn/ggxx/201705/P0201705274933008 72669.pdf> (accessed 28 June 2017).

[7] Shengwu Tansuo, "Interview with Tong Youzhi, Ceo of Suzhou Kintor Pharmaceuticals, Inc.: A Sense of 'Urgency' in Developing New Drugs in China, Mostly to Fill the Gaps" ("zhuan fang kai tuo yao ye zong cai tong you zhi: zhong guo xin yao yan fa "jin po gan" qiang, duo shu wei tian bu kong bai"), biodiscovery.com website, 25 May 2015, <http://www.biodiscover.com/news/celebrity/175802.html> (accessed 16 July 2017).

The Biopharmaceutical Industry

The reported sales of biopharmaceutical companies in SIP in 2016 were 47 billion yuan, up from the 38 billion yuan in 2015. The sector has registered growth of 25–35% in recent years.[8] These enterprises focus on producing cost-effective biosimilar biopharmaceutical drugs which are generic versions of the original biopharmaceutical drugs. As a result, the development time of these drugs is shorter, involving less risks as compared to the original biopharmaceutical drug development.[9]

With success in the development of biosimilar drugs by key players such as Innovent Biologics and CStone Pharmaceuticals, the SIP has moved into more basic science research by setting up joint research laboratories with renowned medical research organisations (Box 2).

Developing biosimilar pharmaceuticals is also a priority at the national level. This is in line with the Chinese government's longstanding policy of promoting cheaper versions of effective drugs while expanding the existing 95% coverage rate to various public health insurance plans for its citizens.[10]

China's pharmaceutical market is markedly different from those in the developed world. In 2015, the market share of generic drugs in China was 64% and 22% for patented drugs, whereas it was 21% and 70% respectively in the United States. In addition, the per capita spending on pharmaceuticals in the United States in 2015 was US$1,036 compared to only US$78 in China.[11]

---

[8] Interview with Yang Zhiping, chairman SIPAC, 7 November 2017.

[9] Biosimilar drugs are generic versions of ordinary small molecule pharmaceuticals. Developing biosimilar drug is easier than developing original biopharmaceuticals.

[10] "Benjamin Shobert Testifies before the U.S.-China Economic and Security Review Commission", The National Bureau of Asian Research, 16 March 2017, <http://www.nbr.org/research/ activity.aspx?id=750> (accessed 17 March 2017).

[11] International Trade Administration, "2016 Top Markets Report Pharmaceuticals Country Case Study, China", <http://trade.gov/topmarkets/pdf/Pharmaceuticals_China.pdf> (accessed 12 June 2017).

Box 2. Development of Biopharmaceutical Industry in SIP

SIP started the development of biopharmaceutical industry in 2006. The dedicated biopharmaceutical incubator arm, bioBay, was set up in 2007. Start-ups came in around the 2010s and by end 2016, bioBay had more than 500 biopharmaceutical enterprises, and employed close to 30,000 employees and more than 60 returnee scientists under the national "Thousand Talents Programme".

SIP seeks to attract basic science-oriented laboratories associated with international universities in order to set up facilities at bioBay. In 2017, new entrants will include Oxford University's Oxford-Suzhou Centre for Advanced Research (OSCAR) and Harvard University's Harvard Weitz Innovation Hub specialising in biopharmaceuticals.

Source: News at bioBay, <http://www.biobay.com.cn/main/index.asp>; "Oxford University Leaders Negotiate Innovation Community Research Partnerships in China", 23 March 2017, <http://www.ox.ac.uk/news/2017-03-23-oxford-university-leaders-negotiate-innovation-community-research-partnerships-china>; "Harvard Weitz Innovation Hub Settles in SIP", SIPAC website, January 2017, <http://www.sipac.gov.cn/english/news/201701/t20170119_524964.htm> (accessed 17 July 2017); "Billions to Support New Leap in Biopharmaceutical Industry" (bai yi zi jin zhu tui sheng wu yi yao chan ye xin kua yue), June 2017,<http://news.sipac.gov.cn/sipnews/yqzt/yqzt2017/201706swyy/xgbd/201706/t20170613_573265.htm> (accessed 17 July 2017).

The extent of support that SIPAC provides to a company it seeks to promote and the close working relationship between SIPAC and that company are best manifested in the case of Innovent Biologics, the acknowledged local champion of the biopharmaceutical industry (Box 3 and Appendix 2).[12]

---

[12] "Economy for 30 Minutes: 'Chinese Speed' for New Anti-Cancer Drugs" (jing ji ban xiao shi kang ai xin yao di "zhong guo su du), Youtube, 27 April 2017, <https://www.youtube.com/watch?v =vkYVqlEyrcs> (accessed 28 April 2017).

Box 3. SIPAC Support for Innovent Biologics

Innovent Biologics is an antibody-focused biopharmaceutical company incepted in August 2011 in the SIP. It owns two production lines of 1,000 litres fermentation capacity each and is currently building four production lines, each with 2,000 litres capacity. SIP earlier provided US$140 million to cover land and initial manufacturing facility costs. The facilities were provided initially rent-free to Innovent on a long-term lease contract. They were the first US Food and Drug Administration (USFDA)-certified Good Manufacturing Practice (GMP) biologics facility in China when it was set up. Innovent Biologics acquired the facilities in 2015 from SIP.

The Chinese Food and Drug Administration (CFDA) promulgated rules on China's biosimilar products in early 2015. The rules spell out standards for comparability of biologics to originator reference drugs. Prior to 2015, none of the 96 biologics approved in China was based on this type of head-to-head comparative study.

Innovent Biologics has been conducting R&D on this type of head-to-head comparative study approach since its founding in 2011. Innovent works closely with the government on updating the CFDA rules on biologics and its practices adhere to the best industry practice worldwide. In 2015, Innovent inked two groundbreaking deals with Eli Lilly and Co of Indianapolis to co-develop and co-commercialise six antibody drugs over 10 years, potentially bringing to Innovent more than US$3 billion in upfront and milestone payments. Innovent will develop, manufacture and commercialise the drugs in China, while Lilly will do the same outside of China. Lilly will pay sales' royalties and foot other payments if the drugs are commercialised outside of China.

The product pipeline and the agreement with Lilly are quite revealing. The company prioritises the local market over the more risky and lucrative international market. Innovent's case highlights the close cooperation between the local government and businesses.

Source: Innovent website, <www.innoventbio.com/en/> (accessed 28 April 2017).

Table 1. World's Top 10 Pharmaceutical Companies' Fund Raising Activities (2016)

Ranking	Company	Type[13]	Amount-US$	Country
1	Moderna Therapeutics	Equity Financing	474 million	US
2	**Innovent Biologics**	**Round D**	**260 million**	**China (SIP)**
3	BlueRock Therapeutics	Round A	225 million	Canada
4	Human Longevity	Round B	220 million	US
5	Intarcia Therapeutics	Equity Financing	215 million	US
6	**CStone Pharmaceuticals**	**Round A**	**150 million**	**China (SIP)**
7	Zymergen	Round B	130 million	US
8	Denati Therapeutics	Round B	130 million	US
9	Unity Biotechnology	Round B	116 million	US
10	Dalcor Pharmaceutical	Round B	110 million	Canada

Source: ReadOne, "Gen: Two of China's Biopharmaceutical Companies Listed among the World's Top Ten for Fund Raising Activities" (GEN: liang jia zhong guo sheng wu yi yao qi ye ru xuan 2016 nian quan qiu rong zi Top10), 21 December 2016, <https://read01.com/8E6R5O.html> (accessed 30 March 2017).

In 2016, two of the world's top 10 private equity fund raising activities in biopharmaceuticals came from SIP: the Innovent Biologics and CStone Pharmaceuticals. Accordingly, SIP has now become one of the world's acknowledged biopharmaceutical development clusters (Table 1).[14]

According to the general manager of Lilly Suzhou Manufacturing, the support rendered by SIPAC and the Suzhou branch of Chinese Food and Drug Administration (CFDA) in their lobbying for the inclusion of the company's product in the annual Jiangsu drug purchase list had contributed to their final decision to continue investing in SIP (see Appendix 3 for Lilly's company profile).

[13] A venture round is a type of funding round used for venture capital financing by which start-up companies obtain investments, generally from venture capitalists and other institutional investors. Round A refers to the first round. B refers to the second round and so on.

[14] NUS and IMCB signed a cooperation agreement with SIPAC in February 2017 to commercialise the former's laboratory findings.

He also explained that his company was subject to stringent hygiene standards set by US Food and Drug Administration (USFDA) and CFDA. Lilly's SIP compound that houses both plant I and plant II had passed the demanding validation requirements and certification procedures and its staff at these two plants have become familiar with the validation and certification procedures. This is a key reason behind the company's decision to spend two billion yuan to build plant III at SIP.[15]

He also added that Lilly (Suzhou) would like to build up local human resources capability to meet its growing needs. It seeks to develop partnerships with universities and research institutes in SIP to address specific problems related to the pharmaceutical industry and provide a source of talent for Lilly's further expansion in SIP.

A senior R&D manager of Qiagen (Suzhou) Translational Medicine Co (who was previously a research scientist with A*Star Singapore) commented that SIP offers a more conducive environment for biomedical research as compared to Singapore. In SIP, she has ready access to a large pool of tissue samples for experimentation and such samples can be obtained within a short span of a few weeks. This contrasts with a delay of months or even more than a year for a much smaller sample collection in Singapore (see Appendix 4 for Qiagen's company profile).[16]

She also observed that in SIP, there was closer collaboration among hospitals, research organisations and other stakeholders which has helped to enhance research efficiency. While her work in SIP is more on clinical testing and application type of work as stipulated by the biopharmaceutical industry, her stint in Singapore was mainly theoretical and academically oriented with little application focus.

The Nanotechnology Industry

Nanotechnology research is one of the most active branches of material science research in the world today. In recent years, China has emerged

---

[15] Interview with the general manager of Lilly (Suzhou) Manufacturing Co Ltd, 9 November 2016.

[16] Interview with a senior R&D manager at Qiagen (Suzhou) Translational Medicine Co Ltd, 10 November 2016.

as a powerhouse in material science research. According to *Nature*, mainland China published more than 30,000 material science articles in 2015, much more than the over 12,000 articles of the United States (in second place) and about 5,000 articles of Japan (in third place).[17] The nanotechnology industry in SIP is riding on this strength of China's R&D support for its future development.

The total value of the nanotechnology industry in SIP in 2016 amounted to 38 billion yuan, growing more than seven times from its initial 4.5 billion yuan in 2011. SIPAC has focused on growing the following four main areas: micro-nano manufacturing, new nano material, nano energy and clean technology, and bio-nano technology (Box 4).

According to Dr Mihail Roco, the founding chair of the US National Science and Technology Council subcommittee on Nanoscale Science, Engineering and Technology, SIP is one of the top eight nanotechnology development clusters in the world.[18]

The Chinese Academy of Sciences set up the Suzhou Institute of Nano-Tech and Nano-Bionics (SINANO) in SIP in the early 2010s, with more than 1,000 scientists and engineers on its payroll. More than half of them are PhD degree holders. SINANO focuses on both basic and applied research, while the start-up nanotechnology companies work on applied research.

Nanotechnology among the three high-tech industries secured the most start-up research funding from the Chinese government. By 2015, the nanotechnology-related businesses and research institutions in SIP had worked on 718 nanotechnology-related research projects with 834 million yuan in grants from the national government.[19]

---

[17] "China's Blue-chip Future", Nature.com website, 24 May 2017, <https://www.nature.com/ nature/journal/v545/n7655_supp/full/545S54a.html> (accessed 29 June 2017).

[18] The other seven are Hsinchu Industrial Technology Research Institute in Taiwan, IMEC/Aachen/Eindhoven/EU Triangle in Belgium/Netherlands, Albany Nanotechnology Centre in the United States, Grenible Nano-Micro Centre in France, Silicon Saxony in Germany, Daejeon Science Park in South Korea and Tsukuba in Japan.

[19] In general, new product development cycle of the nanotechnology sector is considerably shorter than the biopharmaceutical sector. A normal development cycle is 3–5 years on basic

Box 4. Development of Nanotechnology Industry in SIP

The Nanopolis was set up in September 2010 to coordinate the activities of eight key stakeholders in the nanotechnology ecology: universities, R&D institutions, nanotechnology companies, traditional industries, the government, industry service providers, SOE VC funds and private VC. The Nanopolis provides common experimental and manufacturing facilities for start-up enterprises in this field.

SIP adopted the business model of government-led + SOE VC funding + market-based + industrial ecosystem build-up in the development of its nanotechnology sector. The roles of SIP extend beyond the traditional incubator manager. As at end 2016, SIP had attracted 23 nanotechnology-related higher learning institutions, 51 public and private nanotechnology research institutions and more than 400 operating businesses employing over 20,000 staff in SIP.

There were more than 69 "Thousand Talents Programme" returning scientists, 161 scientists under other local government sponsored talent programmes and 15 Chinese academicians working in various projects on nanotechnology as at end 2016.

Source: "Nano Strategy Determines the Future" (na mi zhan lue jue sheng wei lai), Suzhou Nanotechnology Development Fifth Anniversary Report (su zhou na mi 5 zhou nian zhuan kan), 2016, pp. 12–15.

The sales figure and number of companies working in the nanotechnology sector from 2011 to 2016 attest to the rapid build-up of the industry. In 2015, there were already more than 350 start-ups operating in SIP with total sales of 38 billion yuan (Figure 1). They primarily work on the four main areas of nanotechnology: micro-nano manufacturing, new nano material, nano energy and clean technology, and bio-nanotechnology.

science, 1–2 years on technology development phase, followed by varying time periods at the application development phase and end user commercialisation phase.

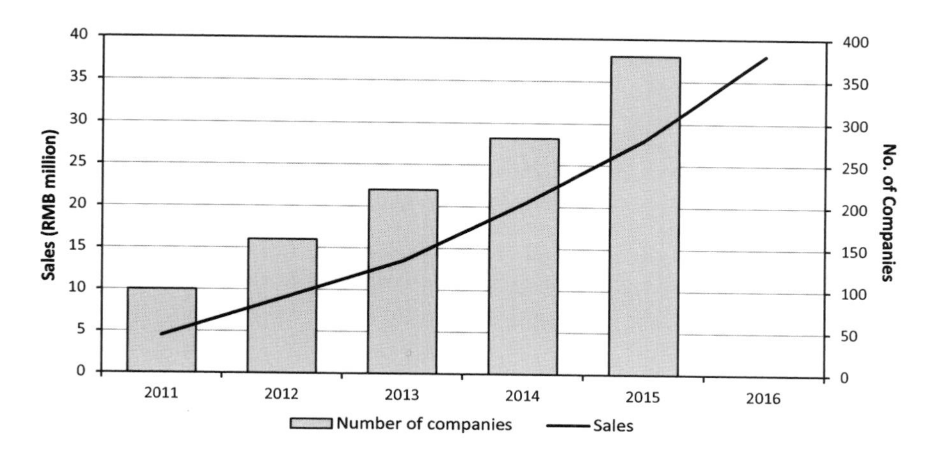

Figure 1. The Nanotechnology Industry in SIP

Note: SIPAC has released the sales of nanotechnology cluster in 2016 but not the number of companies.

Source: "Nano Strategy Determines the Future" (na mi zhan lue jue sheng wei lai), *Suzhou Nanotechnology Development Fifth Anniversary Report* (su zhou na mi 5 zhou nian zhuan kan), 2016, pp. 12–15.

Some start-ups have grown significantly over the years. As of 2015, there were 54 start-ups with annual sales of more than 100 million yuan. Six companies have been listed on the Chinese stock market main board and eight are on the Chinese Growth Board.

The rapid growth in the number of nanotechnology companies and their sales over the past few years, together with the rise of a significant number of companies with hundreds of millions of annual sales, and the localisation of R&D activities thus serve to indicate that the nanotechnology industry cluster has developed to become a viable industry in SIP.

The Cloud Computing Industry

SIP's cloud computing industry[20] in 2016 achieved total sales of 35 billion yuan, up by 25% from 28 billion yuan in 2015. Most of the

[20] Cloud computing is the delivery of computing services: servers, database, networking, software, analytics and more over the internet ("the cloud"). It rose in the mid-late 2000s as internet infrastructure become ubiquitous. Its shared-resource nature generates

Box 5. Development of Cloud Computing Industry in SIP

Cloud computing was one of the three key high-tech industries that SIP promoted in the mid-2000s. This industry drew most of its talents domestically across China and relying less on overseas returnee scientists.

In 2016, SIP had more than 600 information technology companies working on cloud computing with more than 60,000 employees. Among these companies, one is listed on the Shenzhen mainboard and 32 on the National Equities Exchange and Quotations (known as the New Third Board). About 200 of these companies are recognised as leaders in the industry under various government programmes.

Source: "Output Value of Cloud Computing Reaches 35.1 billion RMB", SIPAC, 24 February 2017, <http://www.sipac.gov.cn/english/categoryreport/IndustriesAndEnterprises/201702/t2017 0224_ 534494.htm> (accessed 30 April 2017).

start-up companies in SIP are presently working on software development. Artificial Intelligence was recently added as a new focus area of SIP in April 2017 as SIP has set up a 20 billion yuan investment fund to invest in Artificial Intelligence-related start-ups and businesses (Box 5).

At the national level, cloud computing technology was identified as a key national technology development priority in the Made in China 2025 plan, which is also the Chinese government's fourth Industrial Revolution technology development plan.[21] The emerging Internet of

significant savings for information technology (IT) players. The IT operating facilities are migrating rapidly towards cloud computing worldwide. Three major types of cloud computing services are SaaS, PaaS and IaaS.

[21] Ten Key Sectors for Priority Promotion in "Made in China 2025", <http://www.assolombarda.it/servizi/internazionalizzazione/documenti/made-in-china-2025-key-sectors> (accessed 16 July 2017).

Things (IoT) and data analytics rely heavily on the communication network and high-performance computing component of the cloud computing industry. Cloud computing industry will thus support the next generation of manufacturing industry upgrade under Made in China 2025.

China is set to expand its domestic cloud computing sector.[22] The market for this sector was eight billion yuan in 2013 and by 2020, it is expected to reach 120 billion yuan.[23]

Vice-Minister of Industry and Information Technology Chen Zhaoxiong said in an interview at the Ninth China Cloud Computing Conference held in Beijing in June 2017 that with the fast-evolving cloud computing market, the industry must work on attaining breakthroughs in core key technology and develop new applications in cloud computing to drive the booming digital economy.[24]

The growth prospects of the cloud computing industry thus look very promising. However, competition is especially intense with virtually all major industrial parks across China promoting similar IT industry. Attracting talents and retaining them are thus salient for the industry.

SIP's cloud computing industry used to concentrate on the SaaS segment where local start-ups dominate. However, its composition is slowly changing. Multinationals such as Microsoft had established a R&D centre in SIP while large Chinese companies such as Tong Cheng has built a large cloud computing centre in SIP.

Huawei, China's leading telecommunications equipment giant, is building one of its biggest R&D facilities cum operation centres in SIP, specialising in enterprise and cloud computing. The SIP facility measures

---

[22] "China to Expand Domestic Cloud Computing Sector", *China Daily*, 15 June 2017, <http://www.chinadaily.com.cn/business/tech/2017-06/15/content_29754826.htm> (accessed 20 June 2017).

[23] "2016 Top Markets Report Cloud Computing Country Case Study, China", <http://trade.gov/topmarkets/pdf/Cloud_Computing_China.pdf> (accessed 30 March 2017).

[24] "China to Expand Domestic Cloud Computing Sector", *China Daily*, 15 June 2017, <http://www.chinadaily.com.cn/business/tech/2017-06/15/content_29754826.htm> (accessed 20 June 2017).

759,800 sq m and is expected to house 20,000 research staff by 2020. The company already has a staff strength of more than 2,000.[25]

The forthcoming Huawei facility at SIP will serve as its Enterprise Business Group headquarters and R&D hub.[26] The group provides information and communications technology (ICT) infrastructures and solutions to non-telecommunication corporate customers, including finance, energy, government bodies and public safety. The new Huawei facility will offer all three cloud computing services in SIP: SaaS, PaaS and IaaS.

The entry of such an anchor giant as Huawei will certainly boost the development of cloud computing in SIP with the provision of partner and system integrator support. Cloud computing provides a platform for software development, artificial intelligence, computer security and other budding key technologies in its entire IT ecosystem. As a giant in the sector, Huawei can anchor many small and creative start-ups by acting as their technological mentor, platform provider and customer for services such as analytics, microservices and containers.

Underscoring SIPAC's dedication and meticulousness in growing the cloud computing industry, the CEO of Kmerit (Suzhou) Information Science and Technology Co Ltd said that SIPAC goes through a rigorous process to ensure that it selects the right companies to locate at SIP. In his view, SIPAC has so far done a good job in getting good start-ups for SIP.

He further took the view that the companies operating in SIP already offer nationally competitive wages to their IT talents. Together

---

[25] "Lion Awakening: Suzhou Huawei R&D Centre, 'Technology' Sangtian Island the Local Silicon Valley" (xiong shi jue xing, su zhou hua wei yan fa zhong xin shi tan, ke ji sang tian dao zhi jing gui gu), Sohu News, 7 April 2017, <http://www.sohu.com/a/132597938_673544> (accessed 28 June 2017).

[26] Huawei is the biggest telecom equipment manufacturer in the world with 2016 sales of 521.6 billion yuan and net income of 37.1 billion yuan. The company's R&D spending in 2016 was 76.4 billion yuan. The company has three business segments: Carrier Business Group with 2016 sales of 290.6 billion yuan, Consumer Business Group with sales of 179.8 billion yuan and Enterprise Business Group with sales of 40.7 billion yuan. In 2016, the company was the top software provider in China and ranked fifth in cloud centre operations with revenue of two billion yuan. The company has identified enterprise business as a key growth area.

with its good living environment, SIP's competitive salaries should be sufficiently attractive for IT talents to stay and work in SIP.[27]

He also commented that the cloud computing industry in SIP must continue to move up the IT value chain to continue growing. In relation to its competitors, he was of the view that SIP cannot compete against Dalian in terms of lower wage programers, nor can it compete with Guizhou on cheap power to run commodity type data centres.

As one of the least regulated industries in the country, the cloud computing industry in SIP is oriented almost entirely towards China's domestic market. While SIPAC's incentive programmes help convince start-ups to locate in SIP, merely attracting start-ups is not enough to ensure the success of the industry. Ultimately, it is the quality of the start-ups, influx of key industry players such as Huawei and the ability to attract and retain relevant talents that will determine the long-term prospects of the industry.

Moving Forward Despite Challenges

The three high-tech industries generated total sales of 120 billion yuan in 2016; Yang Zhiping expects total sales to exceed 200 billion yuan by 2020. Short- to medium-term prospects of these industries appear good.[28] SIP's model of building high-tech industries is now being studied by other localities in the Jiangsu province.[29]

---

[27] Interview with the CEO and board chairman of Kmerit (Suzhou) Information Science and Technology Co Ltd, 9 November 2016.

[28] "Suzhou Industrial Park: Leading the Way to a More Creative Future", GlobeNewswire, 20 December 2016, <https://globenewswire.com/news-release/2016/12/20/899200/0/en/Suzhou-Industrial-Park-Leading-the-Way-to-a-More-Creative-Future.html> (accessed 30 March 2017).

[29] Jiangsu provincial government issued document #26 in February 2017 asking all Jiangsu cities to learn from the experience of SIP in the following areas: 开放合作 (opening-up the economy), 产业升级 (industrial upgrading), 科技创新 (technological breakthroughs), 体制改革 (administrative reforms) and 城市治理 (city management). See "SIP Getting More Popular upon Provincial Government Document Issuance" (zhe ci yuan qu yao geng hong IB, sheng zheng fu fa liao ge wen, quan sheng du zhu mu liao),

These three industries enjoy strong local government support as they are considered growth sectors for the future. The necessary human capital and technological foundation behind these high-tech industries are based on a combination of favourable factors such as attracting overseas returnees and local talents across China.

In this sense, these three industries are less susceptible to the "footloose" behaviour of tenant companies, a behaviour commonly experienced by many high-tech industrial parks once the business operating environment changes. The three industries can be considered more stable compared to the earlier labour-intensive enterprises that have relocated out of SIP.

The labour requirement of high-tech industries focuses on the upstream R&D side of the value chain and much less on the manufacturing and assembling ends. Future sales per employee can expect to increase with more activities moving from research to production stage. However, future employment in SIP is not expected to grow much even as the production volume of the three high-tech industries continues to grow and replace the lower value-added businesses.

The innovation in high-tech industries is inherently volatile and unpredictable. At present, the three high-tech industries emphasise application type of innovation rather than the basic but disruptive type of innovation. While this focus significantly minimises commercial risks associated with the basic type of R&D, the growth prospects could also be constrained over the longer term. How the three industries will evolve in future needs to be closely monitored, particularly as they go further upscale towards more basic science-oriented R&D.

Despite all these challenges, the three high-tech industries appear to have built up a complete industry chain from upstream R&D to downstream assembly/manufacturing with home-grown talents and returnee scientists. As these three industries are domestic market-oriented and fit into the national industrial policy, these favourable factors should keep them growing in the foreseeable future.

Weixin, 28 February 2016, <http://mp.weixin.qq.com/s/aErOqSJkXPzgjFdBvdDPeA> (accessed 23 March 2017).

Appendix 1. Incentive Programme for Talent and Business in SIP

SIP announced an updated incentive programme in September 2016 to attract both overseas and home-grown Chinese talents to work in SIP. The key features of the incentives are:

1. Home purchase subsidy

   For academicians from the Chinese Academy of Sciences and Chinese Academy of Engineering, developed countries' academicians, national science award winners and Nobel Laureates, SIP will provide a maximum subsidy of five million yuan.

   For major science awardee and major research project team leader, SIP will provide maximum subsidy of 1.2 million yuan.

   For "Thousand Talents Programme" and SIP technology leadership holders, SIP will provide a subsidy of 500,000 to one million yuan.

   For post-doc fellows trained in the post-doc station in SIP, SIP will provide a maximum subsidy of 300,000 yuan.

2. Housing priority

   All degree holders above bachelor's level and classified as workers in shortage by SIPAC will be given priority for a 2–3 year stay in a government apartment.

3. Income tax rebate

   For talents qualified under Park incentive programme and pay income tax in the Park, SIP will give a one-time tax rebate not exceeding one million yuan based on his/her previous three-year income tax payment.

(*Continued*)

(*Continued*)

4. Wage subsidy

   For companies working under SIP pioneer status, employees holding master's degree or higher with annual salary of more than 300,000 yuan can apply for wage subsidy of 30,000 to 50,000 yuan annually; the subsidy is subject to a cap of three years.

5. Training subsidy

   SIP provides subsidies to companies for setting up in-house training programmes and incentives to company employees for attending approved training programmes. The subsidy and incentive is 50% of company spending and subject to a maximum cap.

6. Post-doc subsidy

   SIP extended subsidy for companies to set-up post-doc stations and pay the wages of post-docs working on approved projects.

7. Hiring subsidy

   SIP will subsidise 50% of the recruitment fee of workers earning more than 300,000 yuan annually, subject to a cap of 150,000 yuan per position and one million yuan per company.

8. Internship subsidy

   SIP will subsidise approved company interns at 800 yuan/month for university students and 1,000 yuan/month for

(***Continued***)

master's students. The subsidy is subject to four months/intern and 50 interns per month per company.

9. Medical subsidy

   All approved talents are provided free annual comprehensive health check-up and VIP medical card for hospital visits in SIP.

10. Other incentives

    Include scholarship subsidy to deserving company employees, five-year long-term stay visa for family of foreign talents and foreign exchange convertibility for legitimate income earned in SIP.

Source: "Notes on Preferential Policies on Attracting Skilled Talents and Talents of Shortage to Suzhou Industrial Park by SIPAC" (yuán qū gōng wěi 、 guǎn wěi huì guān yú sū zhōu gōng yè yuán qū xī yǐn gāo céng cì hé jǐn quē rén cái yōu huì zhèng cè dí yì jiàn), SIPAC, <http://sme.sipac.gov.cn/Policy/PolicyDetail.aspx?ContentID=13464> (accessed 22 September 2016).

Appendix 2. Innovent Biologics Inc

Innovent Biologics is an antibody-focused company based in SIP. It was founded in August 2011 by Dr Michael Yu, a returning scientist from the United States with 20 years of research experience in biopharmaceuticals. Dr Yu is lauded in an official article by *Xinhua News Agency* and considered a national science celebrity.

The company focuses on four diagnostic areas: autoimmune disorder, tumour, ophthalmology ocular fundus disease and cardiovascular disease. In a span of five years, the company produced 12 monoclonal antibody (mab), of which four entered different phases of clinical study.

Three of the mabs under phase III clinical studies are biosimilar products of some of the most valuable biologic drugs in the market. Rituximab attained a revenue of US$47.4 billion in 2015. Adalimumab's sale was US$14.4 billion and Bevacizumab's sale was US$6.9 billion. These three biopharmaceuticals were among the most valuable biologics worldwide in 2015.

Company Code	Molecule	Clinical Approval Date	Clinical Test Phase	Announcement Date	Clinical Test Scale (Patient Number)	Bio Similar Drug	Disease
IBI 301	Anti-CD20 mab	2014/9/24	III	2016/8/19	400	Rituxan (Rituximab)	Non-Hodgkin Disease
IBI 303	Anti-TNFα mab	2016/1/13	III	2016/9/19	400	Humira (Adalimumab)	Ankylosing Spondylitis
IBI 308	Anti-PD-1 mab	2016/9/4	I	2016/9/29	104	New drug	Advanced Solid Tumour
IBI 305	Anti-VEGF mab	2016/5/19	III	2016/11/17	436	Avastin (Bevacizumab)	Non-small cell lung cancer

(***Continued***)

The company completed four rounds of private capital fund raising from 2011 to 2016. The first round raised US$5 million from Fidelity in October 2011; the second round US$30 million from Lilly Asia Venture in January 2015; the third round US$115 million from Legend Capital and Temasek; and the fourth round US$260 million from the State Development Investment Corporation (SDIC) VC arm led the institutional investor consortium.

The company employed more than 400 employees at the end of 2016, of which 350 were engaged in R&D and more than one third hold master's and PhD degrees. Forty of the employees were returning scientists from overseas.

Source: Shannon Ellis, "Innovent Biologics Preps Phase III Trials for Mabthera, Humira Biosimilars in China within Four Months", 18 November 2016, <http://www.innoventbio.com/en/News.aspx?key=news&Id=1365&type=%E6%96%B0%E9%97%BB%E4%B8%AD%E5%BF%83> (accessed 5 February 2018).
Xinhua News, "zai xin qi dian shang, yong pan shi jie ke ji gao feng! — luo shi xi jin ping zong shu ji zai "ke ji san hui" shang di zhong yao jiang hua yi zhou nian shu ping", 30 May 2017, Xinhua.com website, <http://news.xinhuanet.com/politics/2017-05/30/c_1121058820.htm> (accessed 17 July 2017).

Appendix 3. Eli Lilly Suzhou Pharmaceutical Co Ltd

Lilly is one of the top 10 global pharmaceutical companies in the world with annual sales of more than US$22 billion. It is also one of the world's top four insulin producers.

Eli Lilly Suzhou Pharmaceutical Co Ltd set up its first insulin manufacturing facility in SIP in 1996 and the second in 2009. It has been building the third insulin manufacturing plant at the erstwhile vacant lot inside the company compound in SIP since 2014 and the plant had been undergoing pre-production testing as of April 2017. This third manufacturing plant will cost two billion yuan and produce the latest long-acting insulin glargine cartridge system.

Lilly China Research and Development Centre (LCRDC) in Shanghai was opened in 2012. The research facility employs more than 150 scientists and it is investigating new diabetes medicine tailored specifically for the Chinese population. Chinese diabetic patients are different from non-Asian patients because they have a significantly lower average body mass index (BMI) and a higher prevalence of abdominal obesity, fatty liver and insulin resistance. The Suzhou manufacturing plants serve as the complementing production arm of the LCRDC.

As of 2015, there were more than 110 million diabetic patients, or eight per cent of the Chinese population in China. The number of diabetic patients in China is the largest in the world, more than three times that of the United States.

Multinational pharmaceutical companies supply 80% of insulin requirement in the domestic Chinese market and Lilly is the second largest supplier of insulin to the Chinese market, just behind Denmark's Novo-Nordisk.

Source: Firstword Pharma, "Eli Lilly to Open a Third Insulin Plant in Suzhou", 30 March 2017, Firstwordpharma.com website, <https://www.firstwordpharma.com/node/1199355?tsid=17>; Xi'an Jiangtong Liverpool University, "Biological Sciences Students Visit Insulin Processing Facility", 21 April 2017, Xi'an Jiangtong Liverpool University website, <http://www.xjtlu.edu.cn/en/ news/2017/04/biological-sciences-students-visit-insulin-processing-facility> (accessed 17 July 2017).

Appendix 4. Qiagen (Suzhou) Translational Medicine Co Ltd

Qiagen (Suzhou) Translational Medicine Co Ltd is the first diagnostic R&D company focusing on translational medical science in China. Translational medical science (TM) refers to an interdisciplinary branch of the biomedical field supported by three main pillars: bench side (research side), bedside (patient side) and community (medical community side). TM combines different disciplines, resources, expertise and techniques within these pillars to promote enhancements in prevention, diagnosis and therapy. It is a rapidly growing field in biomedical research and the multi-disciplinary, highly collaborative "bench-to-bedside" approach is considered an innovative discipline. Academic programme on TM at MSc and PhD levels started only in the 2010s.

The company provides complete solutions to precision medicine through translational science; it specifically provides services on molecular assaying with the QIAGEN® molecular technologies. The company is a contract research organisation serving the needs of biopharmaceutical companies in SIP and the surrounding Yangtze River Delta area. The company serves its customers through four customer classes: Molecular Diagnostics (human health care), Applied Testing (forensics, veterinary testing and food safety), Pharma (pharmaceutical and biotechnology companies) and Academia (life sciences research).

The company was set up in 2013 in SIP as a joint venture between Qiagen NV and China SIP Biotechnology Development Co Ltd. It is headed by an experienced Chinese-returning biomedical entrepreneur, Dr Peizhuo Zhang. Qiagen NV is a NASDAQ-listed company based in Germany. It has 4,600 employees working at more than 35 laboratories worldwide, and at the end of FY 2015, the company generated revenue of US$1,281 million and net profit of US$249.5 million. China SIP Biotechnology Development Co Ltd is an indirectly wholly

(*Continued*)

(*Continued*)

owned subsidiary of SIPAC through its SOEs network. A great majority of the employees in the company are holders of MSc and higher academic qualifications.

Source: QIAGEN (Suzhou) Translational Medicine Co, <https://biopharmadealmakers.nature. com/users/7416-qiagen-suzhou-translational-medicine-co> (accessed 30 March 2017).

Chapter 5

A Distinctive Model of Social Development*

Introduction

SIP is well known for its remarkable industrial development. EAI's fieldwork in April and November 2016 suggests that SIP's achievements in social development are equally remarkable. Starting from farmland and rural settlements in 1994, SIP has evolved into a highly liveable city with excellent educational and health services, clean environment, vibrant communities, manageable traffic and affordable housing, even by some international standards.

SIP's social development has a number of distinctive features. First, its social policy is development-oriented. It has prioritised building good schools, including local and international schools. The most recent example is Overseas Chinese Academy Suzhou set up to attract Chinese talents from overseas. Second, SIP has maintained a good balance between liveability and development. It has vibrant communities, accessible and affordable social services, clean and safe environment, and good transportation. It has largely averted the common urban ills of poverty, pollution, congestion and insecurity. Third, SIP's way of urbanisation rids itself of the many problems of "urban villages" plaguing governments in Beijing, Shanghai, Guangzhou and Shenzhen. SIP's

*This chapter is based on a report that has been released as an *EAI Background Brief*, "Suzhou Industrial Park: Developing a Distinctive Model of Social Development (V)", an EAI team effort. The first draft was prepared by Dr Zhao Litao.

urbanisation is more complete, and its newly urbanised communities are better regulated and integrated.

The fieldwork demonstrates that SIP's strategies and its investment in social development have generated good economic and social returns. Its convenient neighbourhood centres and particularly good schools attract home buyers and boost land and housing sales; its good social amenities and environment attract techno entrepreneurs and top talents. SIP's development experience is likely to gain more attention in China. It has all the essential features of an "innovative, coordinated, green, open and shared" development model (*chuang xin, xie tiao, lu se, kai fang, gong xiang fa zhan*) that China is currently promoting.

Singapore has played an important role in SIP's social development through the transfer of development "software". Singapore's experience of urban planning, development-oriented social policy, commitment to high standards in environmental protection and its neighbourhood centre model have left an indelible imprint on SIP. In the process of adaptation, SIP has also made noteworthy innovations such as the provision of a more integrated and wider range of facilities provided by its neighbourhood centres. Most importantly, it has avoided the problem of "urban villages" plaguing other cities in China.

As SIP continues to develop and evolve, there is growing room for knowledge sharing between SIP and Singapore in their common search for ways to make communities even more vibrant and the city even "smarter" and more liveable.

Remarkable Achievements in Social Development

SIP, the first G-to-G project between Singapore and China, is well known as a successful model of economic development. An often overlooked but no less important aspect of SIP is its success in social development.

EAI research team's 2016 fieldwork in SIP focused as much attention to its state of development and progress towards industrial upgrading as it was on its social development as its economic transformation (see Table 1 for selected indicators of development).

Table 1. A Comparison of Selected Development Indicators on SIP, Suzhou, Jiangsu and National Averages in 2015

	SIP	Suzhou	Jiangsu Province	China
Per capita GDP (yuan)	257,900	136,700	88,000	49,992
Per capita disposable income of urban residents (yuan)	56,696	50,390	37,173	31,195
Energy efficiency (ton standard coal per 10,000 yuan of GDP)	0.26	0.57	0.51	1.59
Innovation (number of patents per 10,000 people)	86	27	14	—

Source: Data for SIP, Suzhou and Jiangsu are from SIP Planning Exhibition Hall; national data are from *China Statistical Yearbook 2016*.

Due to good urban planning, sustained economic growth and sound social policy, SIP has since evolved into one of the highly liveable cities in China, with excellent educational and health services, clean environment, vibrant communities, manageable traffic and affordable housing.

SIP's social development process has a number of features. First, its overall social policy is highly development-oriented. It has given priority to build good schools, both local and international ones. The fieldwork shows that good schools in SIP have attracted homebuyers as well as talents.

Second, SIP has managed to maintain a good balance between liveability and rapid development. In SIP, its environment is safe, community services are easily accessible and affordable, and traffic is largely free from the congestion that is plaguing other cities in China. The liveability of SIP is set to become even more pronounced over time.

Third, SIP has been completely urbanised by transforming its rural villages into urban communities. In the earlier stages of rapid urban expansion in China, many large cities chose to "urbanise" farmland, leaving residential land to rural collectives.[1] This partial urbanisation

[1] "Urban villages" are a result of incomplete urbanisation. Unlike urban enclaves found in Western countries, "urban villages" in China are governed by formal party and government organisations, including the village-level party committee and administrative

approach has created many enclaves known as "urban villages" or "villages in the city" (*cheng zhong cun*), which has posed a lot of problems to the authorities in Beijing, Shanghai, Guangzhou and Shenzhen. SIP's approach has largely avoided the problem of "urban villages".

SIP's social development has also facilitated its economic development and transformation in a number of ways. To begin with, its convenient neighbourhood centres and particularly good schools attract homebuyers from afar, pushing up land and housing sales.[2] The SIP government needs such revenues to fund its industrial upgrading programmes.

In turn, good social amenities and a pleasant living environment have helped SIP to attract talents from overseas and other parts of China, thereby contributing to SIP's drive to become a high-tech and innovation hub. At the same time, SIP has also begun to "go out" with its own development experiences as a transferable "software" to other cities in China.

As a result, SIP's experience in social development is likely to gain attention in China. Many industrial parks and development zones elsewhere are facing the problem of losing their industrial capacities at the lower-end. SIP will be a reference for achieving a net gain in population and net increase in output value via successful industrial restructuring.

From a broader perspective, China is seeking to move from its previous single-minded pursuit of GNPism (featuring single-minded pursuit of gross national product at the cost of environment and other

---

committee. China operates a dual land management system, with urban land belonging to the state and rural land (including farm land, rural construction land and residential land) belonging to the collectives. Unlike urban land which is tradeable, rural land cannot be sold by law. During expansion, many city governments only acquire farmland and rural construction land, leaving the residential components largely intact. The rationale for doing this is to reduce monetary compensation and save costs. If city governments were to also acquire the residential land, they would have to provide alternative housing to villagers who are left without a home.

[2] Interview with the deputy party secretary and deputy general manager of SIP Neighborhood Centre Development Co Ltd, 7 November 2016.

social considerations) to a new model of "innovative", "coordinated", "green", "open" and "shared (or inclusive)" development. In many ways, SIP is the new development model promoted by China's 13th Five-Year Plan (2016–2020).

SIP's experience shows that successful economic development and social development can be mutually reinforcing. In particular, it has demonstrated that investment in social development is also good for economic growth in the long run. SIP therefore represents a "social investment" approach, which is an alternative paradigm to either GNPism or "state welfarism".

Singapore has played an important role in SIP's social development. Apart from its direct involvement in SIP's physical development through the joint venture company, China-Singapore Suzhou Industrial Park Development Co Ltd (CSSD, *zhong xin ji tuan*),[3] Singapore has also transferred its economic and social development "software".

Singapore's experience of urban planning, development-oriented social policy, commitment to high-standard environmental protection and its neighbourhood centre (*lin li zhong xin*) model featuring integrated amenities and one-stop service centres have left an indelible imprint on SIP. Some distinctive "Singapore DNA" in these social development areas are evident.

However, in the process of adaptation, SIP has also made many significant innovations. For instance, it has commercialised neighbourhood centres by creating a "SIP Neighbourhood Centre Development Co Ltd" (*su zhou gong ye yuan qu lin li zhong xin fa zhan you xian gong si*) and registering "neighbourhood centre" as a trademark with China's State Administration for Industry and Commerce. It has also provided public rental housing to attract talents rather than promote homeownership.

Knowledge transfer from Singapore has become an inherent part of SIP's development. As SIP continues to develop and evolve in its social arena, there is growing room for experience sharing between SIP and

---

[3] In June 2009, CSSD was renamed the China-Singapore Suzhou Industrial Park Development Group Co Ltd (while retaining the initials of CSSD).

Singapore as both sides are still striving to add vibrancy to their communities and develop "smarter" and even more liveable cities.

Development-Oriented Social Policy

From the outset, SIP's overall social policy is development-oriented. It serves the purpose of promoting business and attracting talents. SIP as a new industrial city does not have the same welfare burden that has plagued China's old industrial cities, which have a high concentration of retired state-owned enterprise (SOE) workers.

Local *hukou* holders make up less than half of permanent residents in SIP. Official statistics show that as of end 2013, SIP had 1.03 million permanent residents, 40% of whom have a local *hukou* while the rest are migrant workers. Yang Zhiping[4] frankly acknowledged that "migrant workers are not our responsibility; they return to their home town/village if laid off".[5]

SIP's social policy therefore focuses on attracting talents with its wide range of benefits and privileges while providing social services to a smaller number of local *hukou* holders. Migrant workers, despite being the largest group in SIP, are treated as outsiders without equal access to local social services. From SIPAC's perspective, this duality allows for a targeted provision of limited resources and welfare services for its local *hukou* population.

Moreover, the fact that migrant workers are "dispensable" minimises the potentially high restructuring/transition costs, which have plagued many of China's rust-belt cities whereby SOE retirees and laid-off workers as local residents are "non-dispensable" (which facilitated the organisation of massive social protest for compensation or reemployment in the early 2000s).

Our fieldwork indicates that SIP has done a good job in attracting talents from overseas and other parts of China, and providing social

[4] In February 2017, Yang Zhiping was promoted to become the secretary general of Suzhou municipal government. Zhou Xudong replaced Yang Zhiping as deputy secretary of SIP Working Committee and chairman of SIPAC.

[5] Remarks by Yang Zhiping over lunch talk in SIP, 7 November 2016.

services to local *hukou* holders. Insofar as attracting talents is concerned, SIP has set up special schemes for top talents, including leading scientists or researchers and techno entrepreneurs. It also uses wage subsidy and public rental housing to attract more broadly defined talents, namely, university graduates (for more details, see Chapter 6). What really sets SIP apart from other industrial parks and development zones, however, is its good social amenities and good environment.

Interviewees, including techno entrepreneurs, senior managers and R&D professionals, often highlighted that "SIP has good schools". For those with school-age children, they have expressed their satisfaction with SIP's educational system. SIP's strategy of prioritising on building good schools has evidently paid off in terms of attracting talents.

Building good local schools

SIP did not have good schools in the very beginning. If anything, "SIP only had backward rural schools".[6] However, SIPAC did not put off building good schools till the SIP's economy took off; it began building good schools from scratch (see Box 1 for more information).

During the team's visit to Xinghai Primary School, team members were impressed by the active class participation and the wide range of learning activities and opportunities the school offers (Photos 1 and 2).

Setting up international schools

Apart from building good local schools, another strategy that SIP pursued from the very beginning was the setting up of international schools for the children of expatriates working for foreign-invested enterprises/companies. Suzhou Singapore International School (*su zhou xin jia po guo ji xue xiao*, SSIS) was the first to operate in SIP. The team visited SSIS and Dulwich College Suzhou (*su zhou de wei ying guo guo ji xue xiao*), another well-known international school in SIP (see Box 2 for more information).

---

[6] Interview with the head of Education Bureau, Suzhou Industrial Park Administrative Committee, 7 November 2016.

Box 1. SIP's Experience in Building Good Schools

The principal of Xinghai Primary School (*xing hai xiao xue*), recalled that the school had to set up booths to recruit students. Even so, "for the first batch, we had only 20 plus students every class". SIPAC gave priority funding to education and recruited the best possible teachers from other parts of China. Starting from scratch, Xinghai Primary School has now become one of the top schools not just in SIP, but also Suzhou.

The head of SIPAC's Education Bureau said, "SIP's idea is that if we need to set up a new school, we will make it a very good one from the very beginning". Starting from an "educational lowland", SIP in its 22 years of development, has now become a "highland" not just in Suzhou, but also in Jiangsu".

According to Shen, head of SIPAC's Education Bureau, SIP's educational development was guided by four principles: (i) equalisation — SIP's educational system is fully integrated without the rural-urban distinction; (ii) modernisation — SIP has set up standards in facilities, personnel and administration for every school to meet; (iii) diversification — schools in SIP are encouraged to develop their own strengths; and (iv) internationalisation — there are regular exchanges through immersion programmes between SIP schools and schools in Singapore, Canada, the United States and Europe.

SIP has streamlined the administration of its education. Elsewhere in China, "schools are subject to two levels of administration, by the district government and the street/township government, which makes it difficult for the education bureau to coordinate and often leads to unequal development across schools", while in SIP, "after meeting the standards, schools are directly under the administration of SIPAC, which makes it much easier to transfer principals and teachers to where they are needed most".

Source: Interview with the principal of Xinghai Primary School, 7 November 2016; and interview with the head of SIPAC's Education Bureau, 7 November 2016.

Photo 1. Emphasis on Hands-on Project Work by Students in Xinghai Primary School

Source: Taken by Dr Zhao Litao, 8 November 2016.

Photo 2. Children in a Classroom in Xinghai Primary School

Source: Taken by Dr Zhao Litao, 8 November 2016.

Box 2. Suzhou Singapore International School and Dulwich College Suzhou

SSIS

Founded in1996, SSIS initially offered a Singapore-based curriculum as most of the students then were from Singapore. Today, it is a fully authorised International Baccalaureate (IB) World School catering for 2-18 year olds. It offers the IB Primary Years Programmes for students from Nursery to Grade 5, Middle Years Programme for students from Grades 6 to 10, and Diploma Programme for students in Grades 11 and 12, along with a German curriculum for Grades 1 to 4 students and a North American Style High School Diploma for Grades 9 to 12. Currently at about 1,100, its total enrolment has been on a decline.

Dulwich College Suzhou

Founded in Suzhou in 2007, Dulwich College Suzhou is a member of Dulwich College International (DCI), which has over 7,200 students in eight cities (London, Shanghai, Beijing, Seoul, Suzhou, Zhuhai, Singapore and Yangon). Each of the colleges offers an education based on the English National Curriculum (ENC), enhanced and adapted for international needs. Dulwich College Suzhou is divided into three schools, each having its own building with a number of shared whole college facilities: Ducks-Foundation and Key Stage 1, for early learning years 1 to 2/ages 2 to 7, Junior School-Key Stage 2, for years 3 to 6/ages 7 to 11, and Senior School-Key Stages 3, 4 and 5, for years 7 to 13/ages 11 to 18. The total enrolment of Dulwich College Suzhou is about 900.

Source: Brochures titled "Suzhou Singapore International School" and "Dulwich College Suzhou: Information for Parents 2016–17"; enrolment data from interviews done from 8 to 9 November 2016.

SSIS is owned by CSSD. As the lead developer of the 80 sq km Singapore-China Cooperation Zone in SIP, CSSD is in a good position to actively and swiftly respond to the educational needs of expatriates. With CSSD's good connections to SIPAC, SSIS has preferential access to policy benefits SIPAC could offer.

There is mutual dependence between international schools and children of expatriates. International schools are still important for these children in SIP; however, as the number of expatriates is on the decline, international schools are facing mounting challenges in maintaining a stable student enrolment. Economic restructuring with declining FDI and expatriates has adversely affected the intake of the international schools (see Box 3).

Box 3. Challenges Facing International Schools in SIP

A Singaporean teacher at Dulwich College Suzhou also observed a high turnover among her students. Those who stay for a longer period are from families that "have their own business here, or are bi-racial with one parent being a Chinese national, or overseas returnees with a foreign passport". As "old enterprises leave, and expatriates are being replaced by locals, student enrollment is shrinking. For the first time, four classes of a grade are cut down to three classes, each having 19–20 students".

The head of SSIS confirmed the same trend for his school. The target of student enrolment has been "lowered from the current 1,130 to 900 for the near future" as the number of expatriates in SIP is decreasing. Moreover, as companies "cut costs, long-term contracts for expatriates are being replaced by short-term contracts, and school fees are not part of the package".

The fieldwork shows that Dulwich College Suzhou has been actively trying to cope. However, China's existing regulations do

(*Continued*)

Box 3. (*Continued*)

not work in favour of alternative options. Dulwich College Suzhou had tried to recruit international students (who would study in SIP without the company of their parents), but they had difficulty in getting long-term visa for such students. Glory Goh said, "Currently there are only 7-8 such international students".

A more workable solution for Dulwich College Suzhou is to set up a separate private school for Chinese nationals, an option that SSIS is less keen to undertake. Part of the reason is that though SSIS is owned by CSSD, its management team led by a Western educational professional has enjoyed a high degree of operational autonomy. As running SSIS as an international school rather than a private school ensures a higher level of autonomy, Nicholas Little, head of SSIS, viewed the declining number of expatriates "not as a threat, but something [we] have to adjust". What he meant by "adjust" was a lowering of projected student enrolment from the current 1,130 to 900.

A more important reason is that CSSD, which owns SSIS, is pursuing a dual track approach. Instead of pressuring SSIS to establish a separate programme and register it as a private school, CSSD allows SSIS to concentrate on international students, while establishing a new school — Overseas Chinese Academy Suzhou — next to SSIS to focus on children of overseas returnees.

Source: Interview with a Singaporean teacher at Dulwich College Suzhou, 8 November 2016. The teacher previously taught at the SSIS and moved to Dulwich College Suzhou in 2008 soon after it was established. Interview with the head of SSIS, 9 November 2016.

Establishing Overseas Chinese Academy Suzhou

The challenge facing international schools is not necessarily a challenge for SIP. As the number of expatriates is on the decline, SIP has shifted its priority from expatriates to overseas Chinese returnees. CSSD decided to set up a new school for children of overseas returnees.

CSSD has partnered with Chiway Education Group (*zhong rui jiao yu ji tuan*) to establish Overseas Chinese Academy Suzhou (*su zhou gong ye yuan qu hai gui zi nv xue xiao*, OCAS), the first international school catering for the children of overseas Chinese who return to China and Chinese students who are seeking international education. Children whose parents are SIPAC-recognised "leading talents" will enjoy a 25% discount on the tuition fee (set at 90,000 yuan a year, considerably lower than that of other international schools).[7]

The establishment of OCAS shows that when there is a need for a new school to attract talents, relevant institutions like CSSD will act swiftly to take advantage of new opportunities. SIPAC has been proactive in providing a conducive educational environment to attract talents.

A Good Balance between Liveability and Development

The second distinctive feature of SIP's social development is its ability to maintain a good balance between urban liveability and economic development. Indeed, SIP has succeeded in managing both development and urbanisation to ensure a high level of liveability — vibrant communities, accessible social services, clean and safe environment, good infrastructure including roads and public transport, and so on — while avoiding the urban ills of poverty, pollution, congestion and insecurity.

The fieldwork focused on SIP's way of community building and management, in particular, SIP's adaption of Singapore's neighbourhood centre model and how it is making strides in building smart communities.

---

[7] By design, OCAS provides 15 years of K-12 bi-lingual education, emphasising both Chinese and English as the language of instruction. Within the framework of the IB curriculum, it also incorporates Chinese national curriculum. OCAS plans to take in the first batch of students in 2017. See <http://learning.sohu.com/20161024/n471191172.shtml> (accessed 6 February 2017).

SIP's neighbourhood centre model

Singapore has adopted the "neighbourhood concept", which emphasised self-sufficiency when developing affordable, quality public housing and related facilities for Singaporeans. In the 1970s and 1980s, it reorganised shophouses into a retail hierarchy with regional centres, new town centres, neighbourhood centres and precinct shops. Neighbourhood centres provide a wide range of convenience goods and some personal services to local catchment population.[8]

SIP has borrowed the concept of neighbourhood centre from Singapore and adapted it with some distinctive characteristics. The first notable feature is the establishment of a state-owned company to drive the development of neighbourhood centres (see Box 4). This model has the advantage of having preferential access to urban land at much lower prices.[9]

The second feature is the positioning of the neighbourhood centre as a "community commercial centre" and a "community service centre", integrating a wide range of commercial, cultural, recreational, health-care and educational services.

Compared with neighbourhood centres in Singapore, SIP's neighbourhood centres are more comprehensive in terms of functions and facilities.[10] They offer a much wider range of social services, combining

---

[8] Muhammad Faishal Ibrahim and Lim Fong Peng (2015), "The Development of Neighbourhood Centres in Singapore: From Traditional to Cluster Shopping", *Pacific Rim Property Research Journal*, vol. 11, no. 4, pp. 373–392.

[9] Nonetheless, the initial investment, financed by bank loans, is still high for every neighbourhood centre. The pressure for paying off bank loans is still very much a financial burden. Interview with the marketing director of Linray Investment, 8 November 2016.

[10] The team notes that Singapore's first integrated community and lifestyle hub — Our Tampines Hub — was opened on 6 August 2017. This hub features sports and recreational facilities (including a town square, a 5,000-seat stadium, six rooftop swimming pools, a 30-lane bowling centre and a 1,800-seat sport hall), lifestyle and arts facilities (including Tampines Regional Library, a festive arts theatre and an eco-garden), community facilities (including a small plaza, a small mall with eateries and a supermarket, a family medical clinic and community health centre, a community club and a public service centre for residents to use the services of 12 government agencies).

Box 4. SIP Neighbourhood Centre Development Co Ltd

In November 1997, SIPAC established SIP Neighbourhood Centre Development Co Ltd (*su zhou gong ye yuan qu lin li zhong xin fa zhan you xian gong si*), a subsidiary of Sungent Holding Group (*xin jian yuan kong gu ji tuan*). Since then, it has been the driving force behind the planning, developing and operating of neighbourhood centres in SIP.

In 2000, the SIP Neighbourhood Centre Development Co Ltd registered "neighbourhood centre" as a trademark with the State Administration for Industry and Commerce (SAIC); in 2015, SAIC recognised "neighbourhood centre" as a "China's Famous Trademark" (*zhong guo chi ming shang biao*).

By the end of 2015, SIP Neighbourhood Centre Development Co Ltd had developed 17 neighbourhood centre projects and a larger "Linray Square Project" (*lin rui guang chang xiang mu*). Despite the organisation of many commercial activities, every neighbourhood centre is to provide no less than 45% of its floor space for basic services and public services.

Source: Interview with the marketing director of Linray Investment (*lin rui tou zi*), 8 November 2016.

public services with commercial services, and gearing social services to the size and needs of individual neighbouring community (see Table 2).

The third feature is the promotion of SIP's neighbourhood centre model in other Chinese cities as a commercially viable model. Building on its reputation as China's leading player in planning, developing, marketing and operating of community commercial service platforms, SIP Neighbourhood Centre Development Co Ltd has started to "go out" of

See <http://www.straitstimes.com/singapore/pm-lee-opens-our-tampines-hub-singapores-first-integrated-community-and-lifestyle-hub> (accessed 8 August 2017).

Table 2. Services Provided by Neighbourhood Centres in SIP

		Services Provided			
Project Positioning (*xiang mu ding wei*)	**Floor Space (*xiang mu gui mo*)**	**Basic Services (*ji chu fu wu*)**	**Commercial Services (*shang ye fu wu*)**	**Public Services (*gong yi fu wu*)**	**Public Facilities (*gong jian pei tao*)**
Community commercial centre (*she qu shang ye zhong xin*, see Photos 3 and 4)	30,000–45,000 sq m	Wet market, supermarket, banking service, food stalls, repair shops, laundry, barber shops, drug stores and so on	Child care, elder care, health care, recreation, restaurants, apartments and makerspace	Community club, library, cultural and sports centre and clinic	Social Work Committee and citizen square
Convenient service centre (*bian min fu wu zhong xin*)	15,000–30,000 sq m	Same as above	—	Services provided depends on actual needs	—

Source: Brochure of SIP Neighbourhood Centre Development Co Ltd and interview with Chen Jiang, marketing director of Linray Investment, 8 November 2016.

Photo 3. Fangzhou Neighbourhood Centre (*fang zhou lin li zhong xin*)
Source: Taken by Dr Zhao Litao, 8 November 2016.

Photo 4. Supermarket in Fangzhou Neighbourhood Centre
Source: Taken by Lye Liang Fook, 8 November 2016.

SIP. It adopts a strategy of "asset-light expansion" (*qing zi chan kuo zhang*), focusing on consultancy services in design and planning as well as marketing and management services. Its partners are responsible for raising funds to build neighbourhood centres.[11]

Overall, SIP's neighbourhood centre model works well. It has generated tangible economic and social benefits as home buyers are attracted by neighbourhood centres and the vibrant communities created.[12]

Building "smart communities"

Within neighbourhood centres, SIPAC has set up community clubs (*min zhong lian luo suo*) to provide one-stop administrative and social services as well as multi-purpose venues for local residents, including community service centre (*she qu gong zuo zhan*), elders' home (*le ling sheng huo guan*), children's activity room (*shao er yang guang ba*), health service station (*wei sheng fu wu zhan*), community library (*lin li tu shu guan*), and neighbourhood arts and sports centre (*lin li wen ti zhong xin*) (see Photo 5).

In recent years, SIP has paid particular attention to building "smart communities" (*zhi neng she qu*). Based on mobile technology, big data analytics and cloud computing, SIP created its smart community platform in July 2015, which won a commendation from the Ministry of Civil Affairs (see Box 5 for more information).

Based on the fieldwork, the impression was a SIP that has gone beyond Singapore in building smart communities. SIP's community centres are more developed than their counterparts in Singapore for their use of IT to deliver public services, providing well integrated online and offline platforms and having sufficient IT savvy staff to deliver such services.

---

[11] By the end of 2015, it had over 30 projects with its partners in 18 cities, six provinces with a total floor space of nearly 1.2 million sq m. Interview with the marketing director of Linray Investment, 8 November 2016.

[12] One of the interviewees highlighted that neighbourhood centres generate price premiums for nearby flats. Interview with the deputy party secretary and deputy general manager of SIP Neighbourhood Centre Development Co Ltd, 7 November 2016.

Photo 5. Fangzhou Community Club within Fangzhou Neighbourhood Centre

Source: Taken by Dr Zhao Litao, 8 November 2016.

Box 5. Building "Smart Communities" in SIP

SIP's smart community platform has a number of features and breakthroughs. First, it has contributed to the efficiency of administrating the community services. Its on-site online service terminals allow users to have a one-stop access to 63 types of community services (such as health care and family planning, social security, employment, old-age care, public assistance, ethnic and religious affairs, and volunteering) plus 20 types of information services (such as map, transportation, tourist attraction, home service, house rental, job information and so on). Local residents can also access such services from the comfort of their

(*Continued*)

Box 5. (*Continued*)

homes and check out the status of their requested services online, via the mobile phone, telephone, or visit the community service centres in person.

Second, it has also integrated online and offline service platforms. Community service centres within community clubs serve as the offline service platform, which provides one-stop, all-inclusive services delivered by professional social workers. The community service centres provide an alternative to those who prefer offline access to administrative services.

Third, SIP's smart community platform is connected to the information system of various government agencies. Using their identification card or citizen card, local residents can enjoy the convenience of accessing various services using one card. Through information sharing, previously segregated information systems administered by eight government agencies and offices — employment management system, labour contract system, rural-urban medical insurance system, land for security system, retiree management system, food safety supervision system, low-income verification system and volunteers management system — are now integrated.

Source: Fieldwork in SIP, 8-10 November 2016.

The SIP Model of Urbanisation

The third feature of SIP's social development is its distinctive model of urbanisation. It was Yang Zhiping who suggested to the study team to take a look at SIP's newly urbanised communities which were rural villages not long ago; Yang was visibly proud of SIP's achievements in this area.[13] SIPAC arranged for the team to tour the Jintang Community (*jin tang she qu*) (see Box 6 for more information).

[13] Remarks by Yang over lunch, 7 November 2016.

Box 6. Jintang Community

Jintang Community was formed in 2004 by combining 11 administrative villages (*xing zheng cun*) into a new village. The new village was renamed Jintang Community in 2015. Currently, it occupies an area of 321,500 sq m, with gross floor space totalling 288,800 sq m. It has 124 four-storey blocks, 3,000 housing units and 1,549 households. Based on the statistics obtained, a household on average has two housing units. There are 6,180 permanent residents or local *hukou* holders; by comparison, the number of migrants is much larger at 13,000, more than double the number of local residents.

Jintang Community's governance structure is no different from other urban communities. It has a party committee, which oversees four party branches, 12 party cells and 219 party members, and a residents' committee, which oversees 10 residents' groups, and "mass organisations" (*qun tuan zu zhi*) such as the Communist Youth League, Women's Federation, Militia, and Public Security and Dispute Mediation Office.

There is a service centre for party members, which comprises a management station of migrant party members, a work station for computer-assisted education, a training room, a multi-purpose function room, a reading room, a consultation room and a gymnastics room. There is also a community service centre, occupying a floor space of 400 sq m and comes complete with a one-stop service platform, a police office, a comprehensive social management office, a camera surveillance room, a reference room, a mediation room and so on. The service centre for party members is also housed in the community service centre.

Source: Fieldwork, 10 November 2016.

The research team learned that a major work SIP has done since 2013 is a three-year upgrading project for communities like Jintang. The project targeted 37 communities which were at least five years old (in 2013) and had underdeveloped infrastructure and public facilities. SIP invested a total of 1.78 billion yuan, benefitting 50,000 households.

The project involved upgrading of infrastructure (including roads, roofs, drainage, garbage disposal, street lighting, walls and so on) and public and private facilities such as expanding parking lots (from 0.51 per household to 0.83), repairing meter boxes, building mailboxes (one for each household), and adding convenient stores and supermarkets, gymnastics rooms, activity rooms for seniors, reading rooms, multi-purpose function rooms and so on.

The team did not have the opportunity to interview the residents of Jintang Community. The satisfaction level with their compensation for the loss of farmland and residential land and with the redeveloped community is hence unknown.

On employment, the staff of the community service centre revealed that some residents who were previously farmers with low education are now working as cleaners or taking up low-skilled jobs in the service sector. This is compensated by the urban pension and health-care system, and the extra cash when they rent out the extra rooms in their residence to migrant workers.

Evidently, Jintang Community is quite different from "urban villages" in other cities. From a comparative perspective, SIP's approach can be described as a model of "full urbanisation", which "urbanises" not only farmland, but also residential land, and transforms rural villages into urban communities.

By contrast, "urban villages" in other parts of China represent a model of "incomplete urbanisation". Expanding cities often leapfrog rural settlements. To reduce compensation costs and to save time, city governments acquire only farmland for industrial and commercial uses, leaving the residential components of villages largely intact. Due to collective ownership of residential land, village committees cannot be changed into urban residents' committees, which do not own any land.

As they are governed by village committees, "urban villages" are not administratively integrated into the urban system. It is not the responsibility of urban authorities to upgrade infrastructures in urban villages and provide public services.

Meanwhile, "urban villages" are not subject to formal development control. "Villagers" build high-rise buildings, provide low-cost housing for migrant workers and sustain low-level urban services, workshops and small enterprises.

Concerns over under-provided public services, fire and public health hazards, and various forms of social ills have prompted governments in Beijing, Shanghai, Guangzhou, Shenzhen and other cities to take actions to transform "urban villages" into urban communities.

By then, the government has to pay a much higher price than before because land price has risen and "villagers" would ask for more compensation to demolish high-rise buildings they have built.

SIP's approach of full urbanisation largely avoids the problem of "urban villages" and is less costly. Moreover, after moving local residents from scattered rural settlements to compact four-storey blocks (in the case of Jintang Community), SIP has more disposable land for commercial and industrial uses. Overall, SIP's model of urbanisation is good for both economic growth and social development.

For non-locals, SIP operates a "selective migration policy", which is designed to attract human resources needed for SIP's industrial upgrading while denying many social welfare benefits to the low-skilled workers. This dual approach is undoubtedly unfair to the hundreds of thousands of migrant workers.

Nonetheless, in the eyes of development-oriented SIP officials, it allows SIP to concentrate policy benefits on a much smaller number of talents, professionals and skilled technicians. Equally important, it saves the trouble of redeploying/reemploying low-skilled migrant workers laid off during industrial upgrading.

Chapter 6

Social Development for Industrial Transformation*

Introduction

SIP faces the challenge of plant shut-down, factory relocation, and loss of production capacity and migrant workers like other industrial parks and development zones in China. The study team's fieldwork in 2016 suggests that SIP in recent years is doing quite well in meeting these challenges. It has made notable progress in reinventing itself as a kind of "Suzhou Innovation Park" by attracting top talents and high-tech start-ups. So far, SIP has attracted over 130 top talents listed in China's national "Thousand Talents Scheme" (*qian ren ji hua*), topping other industrial parks and development zones.

What sets SIP apart from its competitors in China in attracting these top talents is its liveable urban environment and its conducive social policies. Good schools, clean environment, affordable housing and Suzhou's history and culture are critical elements. As one interviewee highlighted, talent mobility is often a result of family decision: "Your wife has to like the place before you can move here".

SIP's success in social development has been driven by good urban planning and management, sustained economic growth and sound

---

* This chapter is based on a report that has been released as an *EAI Background Brief*, "Suzhou Industrial Park: Harnessing Social Development for Industrial Transformation (VI)", an EAI team effort. The first draft was prepared by Dr Zhao Litao.

social policy. Much of all these has been drawn from Singapore's past development experiences. Today, SIP is reaping early harvests in its investment in various social policies, which have become a key source of its "comprehensive competitiveness" in terms of developing high-tech industries and high value-added and knowledge-intensive activities.

Still, SIP is facing a number of new challenges. For one, some of SIP's preferential sectors, such as biomedicine, are still small in scale and a solid innovation ecology has yet to fully develop. SIP needs to build on its current momentum and attract a critical mass of high-tech start-ups and R&D centres. For another, SIP has allowed the property market to overheat since the autumn of 2015. If the trend continues, SIP's housing will become less affordable and the city less liveable. This could erode a key source of SIP's competitiveness and reduce its overall attractiveness for oversees returnees and home-grown talents. There is hence much room for SIP to balance short-term gains from the land/property market with long-term returns from high-tech firms and R&D centres. With enhanced liveability, SIP will continue to attract top talents and become one of China's leading innovation hubs.

Social Foundations of Industrial Upgrading

Like other industrial parks and development zones, SIP faces the challenge of factory closures or relocation due to rising wages. The loss of manufacturing capacity was quite serious, at about 20 billion yuan in terms of output value.[1] SIP has been actively seeking to replace lost capacity with new capacity that employs more advanced technology and involves more R&D activities.

Since high-level human resource development is most crucial to the growth of high-tech and innovative industries, SIP's plan of attracting

[1] Remarks by a deputy director of the Economic Development Commission, Suzhou Industrial Park Administrative Committee, during our group discussion with SIP officials, 7 November 2016.

and nurturing talents is most fundamental to its success in this phase of development.

The study team's fieldwork in SIP in April and November 2016 sought to gather the latest update on SIP's progress in the transition from "Suzhou Industrial Park" to what an SIP official called the "Suzhou Innovation Park"[2] and a glimpse of such transformation at its early phases.

The interviews with government officials, techno entrepreneurs, senior managers and R&D professionals show that SIP is undoubtedly making good progress in attracting top talents and high-tech start-ups, due probably to its head start advantage over other industrial parks and development zones.

Three contributing factors have been identified. The first is SIP's talent scheme, which selects and awards a small group of "leader-type talents" (*ling jun ren cai*) with generous benefits and incentives. The second is SIP's strategic use of additional social policy packages to attract "talents in short supply" (*jin que ren cai*).[3]

The third is SIP's indirect kind of benefits associated with good social amenities and environment. The fieldwork suggests that this factor matters most to SIP's imported talents, more important than the other two factors when they were assessing different places to work and live.

As one of the interviewees highlighted, talent mobility is often a result of family decision: "Your wife has to like the place before you can relocate here".[4] Good schools, clean environment, affordable housing,

---

[2] During the team's group discussion with SIP officials on 7 November 2016, Shen Yan, chief of Science Service Division, Suzhou Dushu Lake Science and Education Innovation District Administrative Committee, caught the team's attention with his use of this term "Suzhou Innovation Park".

[3] Falling into the category of "talents in short supply" are largely university graduates working in select sectors identified by SIPAC for priority development. "Talents in short supply" are much larger in number than "leading talents".

[4] Interview with the general manager of Lilly Suzhou Pharmaceutical Co Ltd, 9 November 2016.

safety and Suzhou's history and culture are among the most mentioned reasons for the interviewees' family migration to SIP.

As other industrial parks and development zones have introduced comparable talent schemes and engaged in the instrumental use of such "social policy incentives" to attract talents, what really sets SIP apart from its competitors is its good social amenities and outstanding living environment.

SIP's success is a result of good urban planning and management, sustained economic growth and sound basic social policy. Today, the earlier investment in social development has obviously paid off. It has become a key source of SIP's "comprehensive competitiveness" when it comes to inter-locality competition in developing its talent-based high-tech industries and high value-added activities.

Based on the interviews, SIP seems to have adopted the strategy of "walking on two legs": one is to attract talents to its preferential sectors; the other is to attract homebuyers who are willing to pay higher housing prices for access to SIP's distinctive social infrastructures (particularly schools) and environment.

The strategy of attracting homebuyers brings in revenue and residents for SIP. However, the fieldwork also discovers that if the aforementioned trends continue, the overheated property market could make SIP's housing increasingly unaffordable and the city less liveable. This would eventually affect SIP's overall "comprehensive competitiveness".

Some of SIP's preferential sectors, such as biomedicine, are still in their nascent development phase. The scale is still relatively small and a solid innovation ecosystem has yet to form. SIP's transition towards a viable innovation hub will depend on whether it can continue to attract a critical mass of high-tech start-ups and R&D centres.

Equally important, as talents matter most for SIP's future success, there is a need for SIP to keep its policy of "walking on these two legs" balanced by reining in the surging housing price.[5] If SIP remains a

---

[5] The surging housing price in SIP since 2015 has attracted media attention. See <http://www.yicai.com/news/5015819.html> (accessed 16 January 2017). Suzhou's average

liveable city with good social services, affordable housing and a clean environment, it will hold good promises in upgrading itself to be one of China's leading innovation hubs.

SIP's Talent Schemes

SIP has identified human resources as the key to its industrial upgrading and economic transformation. It uses talent schemes to attract high-level talents and/or talents in short supply (*gao ceng ci, jin que xing ren cai*). The most important talent scheme is the "Jinji Lake Double Hundred Talents Scheme" (*jin ji hu shuang bai ren cai ji hua*), launched in 2010, with an initial goal to attract/nurture 200 innovative and entrepreneurial leader-type talents (*chuang xin chuang ye ling jun ren cai*) and 200 high-skilled leader-type talents (*gao ji neng ling jun ren cai*) every year between 2010 and 2015.

The "Jinji Lake Double Hundred Talent Scheme" was renewed in 2015. On the basis of attracting/nurturing 200 innovative and entrepreneurial leader-type talents and 200 high-skilled leader-type talents in 2015, the updated scheme plans to increase the number by 20% every year in the next three years (see Box 1 for more details on subcategories of leading talents).

SIP uses a wide range of incentives and support to reward selected top talents. It also has programmes to reward a much larger number of talents needed by SIP, albeit with lower levels of benefits. Depending on which category they belong to, talents have privileged access to housing benefits, income benefits, health-care benefits, *hukou* and education benefits, visa and residential permit, post-doc subsidies, foreign exchange and remittance, intern subsidies and so on. Enterprises/companies hiring talents also receive various forms of support.

---

housing price increased 32.8% from 11,998 to 15,932 yuan per sq m between October 2015 and October 2016. See <http://fdc.fang.com/index/BaiChengIndex.html> (accessed 8 August 2017). The concern is that if the trend continues, apartments in SIP will become increasingly less affordable.

Box 1. The "Jinji Lake Double Hundred Talent Scheme" (2015–2017)

Five Types of Leader-type Talents (*ling jun ren cai*)

(1) Science and Technology (S&T) Leader-type Talents (*ke ji ling jun ren cai*): 100 every year from SIP-listed industries (no fewer than 40 from Nano technology); the start-ups are further classified into leading projects (*ling jun xing xiang mu*), growing projects (*cheng zhang xing xiang mu*), incubating projects (*fu hua xiang mu*) and innovating projects (*chuang xin xiang mu*), with decreasing levels of privilege and support in start-up capital, venture capital investment, loan guarantee and so on.

(2) High-Level Leader-type Talents (*gao ceng ci ling jun ren cai*): 30 every year and selected from the "One Thousand Talents Scheme" (for overseas returnees), "Ten Thousand Talents Scheme" (for home-grown talents) (operated by the Organisation Department of the Chinese Communist Party Central Committee), "One Hundred Talents Scheme" (operated by the Chinese Academy of Sciences), "Yangtze River Scholar Scheme" (operated by the Ministry of Education) and talent schemes operated by Jiangsu provincial government and Suzhou municipal government.

(3) Leader-type Talents in Science and Education (*ke jiao ling jun ren cai*): 20 every year to be recruited overseas and 20 every year to be hired from SIP-based entrepreneurs with the purpose of improving higher education in SIP and encouraging school-enterprise cooperation.

(4) Leader-type Talents in High-End Services (*gao duan fu wu ye ling jun ren cai*): 30 every year from sectors such as finance, logistics, service outsourcing, culture and creative industries, and so on.

Box 1. (*Continued*)

(5) High-Skilled Leader-type Talents (*gao ji neng ling jun ren cai*): 200 every year.

Applicants for the "Jinji Lake Double Hundred Talents Scheme" have to go through a strict screening process. Dong Kunlin, CEO and board chairman of KMERIT (Suzhou) Information Science & Technology Co Ltd, recalled he had to go through several rounds of rigorous interviews. In 2013, SIPAC awarded KMERIT the status of a "leading growing enterprise" (*ling jun cheng zhang qi ye*), the pre-requisite for Dong to qualify as a SIP's "entrepreneurial leader-type talent". Due to KMERIT's outstanding performance, Dong was named a "Gusu leader-type talent" (*gu su ling jun ren cai*) by Suzhou municipal government in 2015 and a "high-level entrepreneurial talent" (*gao ceng ci chuang ye ren ca*i) by the Jiangsu provincial government.

Source: Fieldwork in SIP, 8–11 November 2016.

Housing benefits

SIP provides housing grant to a few highly selective categories of talents. The most generous package — up to five million yuan — is offered to scientists and engineers with the highest standing, including members of the Chinese Academy of Sciences and the Chinese Academy of Engineering, or their equivalents in foreign countries, recipients of the State Highest Science and Technology Award and Nobel Laureates.

A smaller amount of housing grant — between half and one million yuan — is provided to those under the "Thousand Talents Scheme" (for overseas returnees and foreign talents) and the "Ten Thousand Talents Scheme" (for homegrown talents). Those selected by SIP as S&T Leading Talents enjoy the same privileges. Postdocs who work in SIP after their stints at a SIP-based postdoctoral workstation qualify for a housing grant of up to 300,000 yuan.

Income benefits

SIP gives a one-time reward — up to one million yuan — to those selected by the "Jinji Lake Double Hundred Talents Scheme" or talent schemes run by the central, provincial and Suzhou municipal governments.

Health-care benefits

Designated hospitals in SIP issues VIP health service cards to a highly selective group of talents — academicians, those covered by talent schemes managed by the central, Jiangsu provincial and Suzhou municipal governments and SIP-recognised leading talents except for the category of High-Skilled Leading Talents (see Box 1 for the classification of leading talents). Card holders enjoy a free health screening every year, and VIP treatment to outpatient/emergency services, hospitalisation, home visit and expert consultation.

Hukou and education benefits

SIP-imported "high-level talents" and "talents in short supply" have the privilege of moving the *hukou* or household registration of their spouse and children to the SIP.

Children of senior managers from Hong Kong, Macau and Taiwan can study in local public schools, enjoying the same rights as local residents.

Upon approval, children of leading talents recognised by SIPAC (except High-Skilled Leading Talents) or higher-level government, regardless of their *hukou* status, can go to local public schools. The same privilege is extended to children whose parent holds a PhD degree or is an overseas returnee with a degree from a foreign university.

Visa and residential permit

Foreign passport holders who are leader-type talents as recognised by SIP or higher-level government can change their visa category to talent visa (*ren cai qian zheng*), or obtain a five-year residential permit after arrival. Their spouse and dependent children are eligible for a five-year residential permit.

Foreign exchange and remittance

Overseas returnees who have acquired permanent residency in other countries can convert their renminbi-denominated income into foreign currency and remit through designated banks.

Relocation services

SIP has set up a special agency to provide one-stop services for imported talents, assisting them in applying for *hukou*, renting an apartment, enrolling their children in local public schools and so on.

Social Policy for "Comprehensive Competitiveness"

SIP's talent scheme — the "Jinji Lake Double Hundred Talents Scheme" — covers only a small number of top-notched talents. For a benefit package as generous as the one described earlier, SIP has neither the resources nor the intention to extend it beyond the highly selective "leading talents".

SIPAC understands that SIP needs not only top talents, but also a much larger pool of R&D professionals to stay competitive and viable. To boost SIP's "comprehensive competitiveness" (*zong he jing zheng li*), SIPAC has strategically used social policy as a broader-based talent scheme for R&D professionals.

Such a broader-based talent scheme offers two types of benefit: wage subsidy and public rental housing. Unlike the strict screening process to select "leading talents", the qualification for wage subsidy and public rental housing is linked to the educational level of the applicants.

Wage subsidy

The amount of wage subsidy or living allowance varies with the educational level. More generous support goes to postdocs. Those with a master's degree are entitled to a smaller amount of wage subsidy. Living allowances are also extended to interns from China's elite universities.[6]

Public rental housing

SIPAC has also strategically used public rental housing to boost its "comprehensive competitiveness". Public rental housing by design should be provided to low income households who cannot afford commercial housing. It has become the dominant form of public housing since the 12th Five-Year Plan (2011–2015), which proposed to build 36 million housing units for 20% of China's urban population by 2015.

SIP did not exactly follow the national guideline. Instead, it used public rental housing to attract university graduates (see Box 2 for details of public housing in SIP).

Social Development Matters

To understand the prospect of SIP in industrial upgrading and economic transformation, entrepreneurs, senior managers and R&D professionals interviewed were asked of their choice of the SIP.

While interviewees acknowledged SIP's pro-business environment, efficient governance and comprehensive government support from heavily subsidised land or office space, start-up capital, subsidies to headhunter fees, hiring costs and training costs, to various forms of benefits/privileges for leading talents and university graduates

[6] Based on information collected from the fieldwork, postdocs working in SIP-based postdoctoral workstations qualify for an annual living allowance of 120,000 yuan. Upon approval, those with a master's degree receive an annual wage allowance of 30,000–50,000 yuan for up to three years if they work in high-tech companies, modern service businesses, higher education institutions, research institutes or medical organisations on SIPAC's preferential list. Their Housing Provident Fund contribution is pegged to their real wage rather than capped at a lower level. For interns from China's elite universities, living allowance is capped at 1,000 yuan a month for master's students and up to 800 yuan a month for undergraduate students. Every employment organisation can have up to 50 interns in any given month and subsidy for every intern should not exceed four months.

Box 2. Public Housing in SIP

SIPAC set up SIP Preferential Public Housing Management Company (*su zhou gong ye yuan qu you zu fang guan li gong si*) in January 2008. To show "compliance" with the national guideline, SIPAC renamed the company SIP Public Rental Housing Management Company (*su zhou gong ye yuan qu gong zu fang guan li gong si*, SIP-PRHMC) in November 2015 without changing its orientation and operation.

Land for public rental housing is provided free by SIPAC. SIP-PRHMC is responsible for raising fund for property development on behalf of SIPAC, while construction is done by SIP's own construction companies. Professional property management companies are selected and overseen by SIP-PRHMC to provide daily management services.

So far, SIP has developed nine communities of preferential or public rental housing, with an area of 850,000 sq m and the capacity to accommodate 18,000 people.

Employees can apply through their SIP-based employing organisation if they meet the following requirements: (1) they do not own any housing in Suzhou; (2) they are university graduates, or college graduates whose specialty is on SIP's list of talents in short supply; and (3) they have signed a labour contract with their employing organisation and pay housing provident fund. Priority is given to those who have graduated within the past five years or those with a bachelor's or higher-level degree. SIPAC's government-guided price for public rental housing is around two thirds of the market level. The main housing types include shared apartment for singles (*dan shen he zu xing*) and apartment for families (*jian dan jia ting hu*). The "Jingying Community" (*jing ying gong yu xiao qu*) offers a monthly rental of 800 yuan (including property management fee) for a shared apartment and 1,600 yuan for a family apartment.

(*Continued*)

Box 2. (*Continued*)

Residents in the "Jingying Community" are encouraged to organise hobby groups and community activities. The team learned that activities organised for single males and females to meet and socialise were welcomed by residents there. WeChat (*wei xin*) has made it much easier to organise hobby groups and mobilise residents for community activities.

Residents can use their housing provident fund to pay their monthly rental. For a shared apartment with a monthly rental of 800 yuan, the out-of-pocket payment can be as low as 300 yuan a month.

SIPAC has set up a time limit for beneficiaries of the public rental housing programme. After 2–3 years of stay in one of the nine public rental housing communities, residents are encouraged to purchase their own housing in SIP or they have to consider renting commercial housing.

Source: Fieldwork in SIP, 8–11 November 2016.

belonging to the category of talents in short supply, it is SIP's social amenities and environment that chiefly account for their move to the SIP (see Boxes 3 and 4).[7]

SIP's Attractiveness to Talents

SIP is evidently making good progress in attracting talents and building capacity in nanotechnology and biomedicine. SIP's experience in

[7] When asked, nearly all of the interviewees — mostly techno entrepreneurs, senior managers and prominent researchers who belong to the category of "leading talents" — emphasised SIP's remarkable social development and good environment as what prompted their family migration to SIP. As emphasised by the general manager of Lilly Suzhou Pharmaceutical Co Ltd, "Your wife has to like the place before you can move here".

Box 3. Why Domestic Talents are Attracted to SIP?

Case 1: Interview with the CEO and board chairman of KMERIT (Suzhou) Information Science and Technology Co. Ltd (*kai mei rui de (su zhou) xin xi ke ji gu fen you xian gong si*) on 9 November 2016.

The CEO and board chairman said that SIP or Suzhou "has good mountains, good water, good traffic, and good culture and environment, but most importantly, it has good education for children". He also highlighted that there are community libraries in neighbourhood centres, and one can borrow a book from one neighbourhood centre and return it at another library in a different neighbourhood centre. Such conveniences make the neighbourhood centres in SIP stand out from neighbourhood centres in other cities in China.

KMERIT started in 2008, providing professional IT services to institutional clients of China's treasury and capital market. Its headquarters, together with the R&D centre, is located in SIP, while maintaining wholly owned subsidiaries in Beijing, Shanghai and Shenzhen.

He said that "there is a big wave of overseas Chinese returning to China, and SIP is a good place for business start-ups". Compared to Beijing, Shanghai and Shenzhen, SIP is a liveable city with affordable housing and good infrastructure. In particular, he stressed that SIP has good kindergartens and schools: "Public kindergartens in SIP have first-rate facilities, yet only charge 450 yuan a month; elsewhere public kindergartens charge 1,500 yuan a month, while private ones charge 3,000 yuan".

Beijing, Shanghai and Shenzhen attract a larger pool of talents. That is why KMERIT maintains wholly owned subsidiaries there to tap into the larger base of human resources. Nonetheless, Dong is cautiously optimistic about the prospect of KMERIT's SIP headquarters and R&D centre. So far, the SIP headquarters has attracted and retained over 60 partners, including a dozen senior ones.

(*Continued*)

Box 3. (*Continued*)

Case 2: The vice president of Soochow Securities Co Ltd (*dong wu zheng quan gu fen you xian gong si*), interviewed on 9 November 2016.

The vice president indicated two advantages SIP has in attracting talents. For one, "SIP is a good brand name outside of China, which helps to attract overseas Chinese talents to return". For another, SIP has good environment and affordable housing. The cost of hiring financial professionals is lower in SIP than in Shanghai due to comparatively lower wages. Therefore, SIP should be attractive to smaller financial institutions.

Nonetheless, he opined that SIP cannot compete with Beijing, Shanghai or Shenzhen in terms of building financial centres: "Beijing has political advantages, Shanghai has a big market, while Shenzhen has flexible mechanisms". Moreover, Beijing, Shanghai and Shenzhen have the power to lower their income tax rate, which is important for high-paying financial professionals. By comparison, SIP does not have this power. Soochow Securities has therefore set up a subsidiary in Shanghai to employ talents who have no plans to move to SIP.

attracting talents is three-fold: First, it has set up a talent scheme offering a wide range of services, benefits and privileges to a highly selective group of "leading talents" (as detailed earlier).

Second, SIPAC has strategically tailored social policy with specific benefits to a much larger number of "talents in short supply", mostly university graduates with a bachelor's or higher degree in SIP's preferential sectors. Wage subsidy and public rental housing are not for the low-income residents in SIP, but for attracting and retaining R&D professionals. In this way, social policy has become an effective economic tool in the implementation of its industrial policy and talent scheme.

Box 4. Why are Foreign Talents Attracted to SIP

Case 1: The general manager of Lilly Suzhou Pharmaceutical Co Ltd and vice president in charge of Lilly China manufacturing, interviewed on 9 November 2016.

The general manager was happy with SIP's "location, infrastructure, government support and living environment". In particular, he was surprised to find that "SIP is cleaner than expected". However, his sole complaint was about the internet in SIP, not in terms of its censorship, but rather its grudgingly low internet speed. This was also a surprising problem to the research team in different parts of SIP.

To cut costs, multinational corporations such as Lilly have reduced the number of foreign employees on long-term contracts. They send in short-term project teams to provide consultation or solutions if needed. Meanwhile, they are focusing on "building local capability" by hiring local talents or graduates from reputable international universities. According to Matt Edwards, general manager of Lilly Suzhou Pharmaceutical Co Ltd, they are not worried about competition from Chinese companies as their competitors are a handful of international pharmaceuticals in China.

Case 2: A researcher at QIAGEN (Suzhou) Translational Medicine Co Ltd (*kai jie (su zhou) zhuan hua yi xue yan jiu you xian gong si*), interviewed on 10 November 2016.

The researcher obtained her PhD degree in biology from the National University of Singapore and worked at A*STAR before she joined her husband in a pharmaceutical company in SIP.

While she moved to SIP for family reasons rather than career consideration, she is also quite satisfied with her life and work there. She cited excellent schools for children, good health-care facilities, clean environment and open culture as factors for moving her family to SIP.

(*Continued*)

Box 4. (*Continued*)

When asked to compare her experience of working at A*STAR in Singapore and at QIAGEN, a private company in SIP that provides integrated solutions for biomarker, companion diagnostics and clinical testing for precision medicine and health care, she pointed out that QIAGEN is much less publication oriented, focusing more on financial viability.

Wang was involved in developing the first made-in-Singapore cancer drug while working at A*STAR. Compared to Singapore, she believed that investment in developing cancer drugs in SIP is likely to produce quicker results and higher returns because for cancer biomarkers, Chinese researchers have the advantage of accessing a much larger pool of samples and in a much shorter period of time.

Third, as repeatedly emphasised earlier, SIP has good social amenities and environment. It has good kindergartens and schools, good hospitals, clean environment, affordable housing and convenient neighbourhood centres. SIP is seen as a liveable city with many attractive features by overseas returnees and homegrown talents.

Other industrial parks and development zones have their own talent schemes. The instrumental use of social policy to attract university graduates is also not unique to SIP. What really sets SIP apart from its competitors in China is how SIP has capitalised on its good social amenities and outstanding environment qualities to achieve this specific objective.

Good social amenities and environment are a result of good urban planning and management, sustained economic growth and sound social policy that prioritise "social investment" and liveability. As most industrial parks elsewhere have largely focused on maximising GDP growth, very few have developed such outstanding social amenities and environment as what SIP has achieved.

Singapore's involvement in SIP — through knowledge transfer in urban planning and providing an inspirational model of urban liveability — has also laid a good foundation for SIP's social development. In fact, many aspects of SIP's social development still carry some distinctive Singaporean characteristics, though SIP has subsequently further developed and innovated on the basic Singapore experience.

So far, SIP has been very successful in attracting talents. A total of 135 individuals have been selected by China's most prestigious talent scheme, the "Thousand Talents Scheme", managed by the Organisation Department of the Central Committee of Chinese Communist Party. In particular, SIP accounts for six per cent of the national total in the entrepreneurial category of the "Thousand Talents Scheme".

SIP has nearly 17,000 foreign talents and 5,000 overseas returnees. Over 500 start-ups are operated by those who have returned after completing their study abroad. SIP has the largest workforce with a college or higher degree among China's development zones, totalling 300,000, or 40% of SIP's working population. In 2015, SIP's R&D spending reached 3.35% of its GDP, considerably higher than the national average of 2.07%.

Meeting Continuing Challenges

In their perennial pursuit of industrial upgrading and economic transformation, industrial parks and development zones in China, including SIP, face the constant challenge of plant shut-down, factory relocation, loss of production capacity and laid-off migrant workers.

Thanks to its good social amenities and environment, SIP is in a better position to meet these challenges. So far its industrial upgrading and economic transformation are on the right track as evidenced by its ability to attract high-tech start-ups, techno entrepreneurs, a diversified set of talents and a large number of R&D professionals.

Nonetheless, SIP also faces three types of challenges when reinventing itself into an "innovation park". First, high-tech industries have high risks of "fast to rise, fast to fall". Due to high mortality rates of high-tech firms, SIPAC's investment in some of these firms may not bear expected results.

Second, SIP has its own disadvantages in developing some of its preferential sectors. In cloud computing, for instance, one of the interviewees pointed out that Dalian (a major city in Liaoning province in northeastern China) has a head start advantage over SIP, while Guiyang (capital city of Guizhou province in southwestern China) has a stronger cost advantage — low electricity tariffs that are ideal for running big-data centres. He admitted that he used the service of a big-data centre based in another city even though the International Science Park Data Centre (*guo ke shu ju zhong xin*) is located nearby in the SIP.[8]

In biomedicine, one interviewee acknowledged that BioBAY (*sheng wu na mi yuan*) is still in the "nascent stage", with over 300 biopharmaceutical companies. The sector is "still small in scale" and "needs more time to grow".[9] Whether SIP can build a critical mass of biopharmaceutical companies is critical for the future success of BioBAY.

In financial services, "talents, information and policy are the three most important factors; SIP cannot compete with Beijing, Shanghai or Shenzhen in each and every dimension". In high-paying sectors such as financial services, "the individual income tax rate matters a great deal".

For financial professionals, "the tax burden is high at 35–40% in China, compared to about 15% in Hong Kong; Shenzhen and Shanghai are able to offer some concession, but SIP does not have the power to do so as the income tax rate for financial professionals is determined by the provincial government".[10]

[8] Interview with the CEO and board chairman of KMERIT (Suzhou) Information Science and Technology Co. Ltd. on 9 November 2016.

[9] Remarks by the chief of Science Service Division, Suzhou Dushu Lake Science and Education Innovation District Administrative Committee, 7 November 2016.

[10] Interview with the vice president of Soochow Securities Co., Ltd., 9 November 2016.

Instead of attracting talents from Shanghai, Soochow Securities Co Ltd, which is headquartered in SIP, found it necessary to establish a subsidiary in Shanghai to tap into the much larger pool of talents there. Other companies interviewed also lamented that "SIP has organised job fairs in local colleges and universities in Suzhou, but we find them not very useful; talents are still concentrated in Beijing and Shanghai".[11]

The third challenge is whether SIP can continue to make its housing affordable and the city highly liveable. Based on interviews, SIP seems to have adopted the strategy of "walking on two legs" by attracting talents to its preferential sectors and attracting homebuyers.

So far, the strategy of attracting homebuyers works quite well. SIP's good social amenities and environment, particularly its good schools, are equally attractive to homebuyers from neighbouring areas and talents from other parts of China. "A lot of residents in the old districts of Suzhou have moved out, with the majority moving to SIP because of good education here; only a small minority moved to the Suzhou New District (*su zhou gao xin qu*)".[12]

Another interviewee confirmed that "in the past, it was the prices of the housing around neighbourhood centres that kept increasing; now it is the prices of the housing around schools".[13]

In the short run, having a large number of homebuyers to move into SIP brings in revenue and population. This strategy — through the revenue it generates — probably makes it possible for SIPAC to implement its other strategy of luring financial institutions and high-tech start-ups with heavily subsidised office towers/spaces and attracting talents through various forms of subsidies and benefits.

This strategy, if not managed well, could easily lead to an overheated property market. In the long run, if housing becomes much less

---

[11] Interview with the CEO and board chairman of KMERIT (Suzhou) Information Science and Technology Co Ltd, 9 November 2016.

[12] Remarks by head of Education Bureau, Suzhou Industrial Park Administrative Committee, 7 November 2016.

[13] Remarks by the deputy party secretary and deputy general manager of SIP Neighbourhood Centre Development Co Ltd, 7 November 2016.

affordable, the foundation of a liveable and sustainable city would be eroded, and one of the most important sources of SIP's "comprehensive competitiveness" would be gone.

This concern has its grounds. The surging housing price has attracted media attention. "The Crazy Suzhou Housing Market" was the title that was used to describe the situation in Suzhou since the autumn of 2015, which was triggered by surging land prices to the east of Jinji Lake in SIP. The media report cited a case in which the price of an apartment was slightly higher than 10,000 yuan per sq m in the summer of 2015. In early 2016, it increased to nearly 30,000 yuan per sq m.[14]

In fact, one SIPAC official interviewed cautioned that the success of SIP's transformation should not be overestimated as "the direction of transition is still not clear". He was critical of what he described as the "securitisation of land and housing", which he believed could create problems for SIP in the long run. The general public's conventional work ethics will be eroded as they "want to make quick money (from the housing market) instead of working hard".

Entrepreneurs would also want to make quick money by "securitising their enterprises through IPOs, transfer of ownership rights and relocation (*dong qian*)". They could profit from the relocation as "they can sell or actually return the land to the government for a good price".[15]

Industrial upgrading and economic transformation are a dynamic process. While making good progress in economic transformation so

---

[14] <http://www.yicai.com/news/5015819.html> (accessed 16 January 2017). Suzhou's average housing price is considerably lower than that of first-tier cities such as Beijing, Shanghai and Shenzhen. Nonetheless, it is higher than that of neighbouring cities such as Wuxi. More notable is its higher growth rate in average housing price. Between October 2015 and October 2016, the average housing price increased 32.8% from 11,998 to 15,932 yuan per sq m. By comparison, Beijing registered a growth rate of 18.4% (from 34,596 to 40,948 yuan per sq m), while Wuxi increased 29.7% (from 8,180 to 10,610 yuan per sq m). See <http://fdc.fang.com/index/BaiChengIndex.html> (accessed 8 August 2017).

[15] The interview was conducted on 7 November 2016.

far, SIP's future success depends on whether it can attract a critical mass of top talents, high-tech start-ups and R&D centres, which in turn hinges on whether it can continue to make work and life in SIP pleasant, affordable, convenient and safe. In this sense, SIP's social development progress underscores its future economic growth and industrial transformation.

Index